© Copyright 2024, Prasad Prakash Tupache.

All rights are reserved. No part of this book may be reproduced or transmitted in any form by any means; electronic or mechanical including photography, recording or any information storage or retrieval system; without the prior written consent of its author.

The opinions/content expressed in this book are solely of author and do not represent the opinion /standings/thoughts **Amazon Kindle Direct** Publication. No responsibility or liability is assumed by the publisher for any injury, damage or financial loss sustained to a person or property by the use of any information in this book, personal or otherwise, directly or indirectly. While every effort has been made to ensure reliability and accuracy of the information within, all liability, negligence or otherwise, by any use, misuse or abuse of the operation of any method, strategy, instruction or idea contained in the material herein is the sole responsibility of the reader .Any copyright not held by the publisher are owned by their respective authors. All information in this book is generalized and presented only for informational purpose "as it is "without warranty or guarantee of any kind.

All trademarks and brands referred to in this book are only for illustrative purpose are property of their respective owners and not affiliated with this publication in any way .The trademarks being used without the permission don't authorize their association or sponsorship with book.

ISBN : 9798336229998

Price

Publishing Year: 2024

Published and Printed By

Independently Published Through Amazon Kindle Direct Publication

Office Address: Amazon (India) Brigade Gateway, 8 Th Floor,
26/1, Dr. Raj Kumar Road , Malleshwaram (W) ,
Bangalore – 560055
Phones: +918033273000
E-mail : amznindpr@amazon.com
Website: www. Amazon.in

Printed in India & Various International Amazon Marketplace (Website) Through Print on Demand Technology

WITH BEST COMPLIMENTS !

M/S TUPACHE CONSULTANTS

PROPRIETOR: MR.PRASAD PRAKASH TUPACHE

UAN: MH26D0030607

SURVEY NO 79/20, SHIV RATNA COLONY,

PACHPIR CHAUK, KOKANE NAGAR, KALEWADI, PIMPRI, PUNE- 411017

CONTACT: 9970173983

ABOUT THE AUTHOR

Author : Mr. Prasad P. Tupache ,

Address: Survey No 79/20,

Shivratna Colony , Pachpir Chauk,

Kokane Nagar, Kalewadi ,

Pimpri, Pune – 411017

Font Setting :   **Publisher** : Amazon KDP

Mr. Prasad P. Tupache .

Cover Design :

Mr. Prasad P. Tupache .

Photo Courtesy :

Front Page : Canva Creative Studio

Rear Page : Canva.com

Created From : Canva.com

BLESSINGS

Image Courtesy: Yogesh Pedamkar, Unsplash.com

Namaste Friends,

Greetings of the season !

With completion of 3 books out of 5 book series of ' The Digital Intelligence , we have decided to start writing an already declared project on thrills observed in Engineering Inspections known as " *A precise commentary on thrilling engineering inspections* - **Check Thrills** – *Learnings of Engineering Inspection Services* "

The other two volumes of 'The Digital Intelligence 'which are based on multiple choice questions based on concepts discussed in the project will be started after completion of this project !

Bappa's blessings always remained as

strength behind every noble thought expressed in this creative cum technical project ! These thoughts are nothing but easy simplification of difficult engineering situations and observations find during close inspection of Engineering Parts , both by manual & Semi-automatic / Fully automatic way of manufacturing !

The importance of every engineering component is well known and as per traditional wisdom of last 200 or more years, it reflects, machines are used to reduce human efforts , increase productivity and fulfill large scale of product or service demand by implementing most 'safe -quick -economical' process of creation ! So , basically it is about providing freedom of choice of doing a work either manually or with the help of automatic or semi-automatic machines !

As far as Quality of any work is concerned, it is recognized by its ' Fitness for Use ' or ' Adherence to laid down specifications, standards and codes ' on the basis of proven performance test , dimensional verification and overall symmetry of the part as per prescribed scale or unscaled drawing !

Writing a report is one thing and writing a quality certified report is another thing ! How ?

The basic purpose of Quality certified report is to present the overall observations find during visual , documentary and test-based processes ! Unless & until there is indication of safe test result as per mentioned technical range of that product , the Quality Engineer is not authorized to sign that test report ! This is where keen observations of Quality Engineer are reflected in front of the overall quality management system ! One can say that the role of Quality Engineers is more like a Judge or an Umpire who observes things , witness the show , verify proofs produced and accordingly gives his decision of Approval – Hold for required Conformance or Reject for major Non - Conformance to the QA system specification laid down ! Through this book , the thrill of inspections will be discussed to ensure correct quality certified certificate ! **Welcome to Check Thrills !** Morya! Morya!

DEDICATION

Image Courtesy: Joanna Kosinska, Unsplash.com

This book is dedicated to truth loving engineers, task master workmen, safest designers, prompt site safety personnel, road sensible transporters, public safety committed high performance management and every customer who uses supplied product as per specified guidelines and instructions!

This book is also dedicated to fighting spirits in any organization who believes in their decisions and never compromise to their expected safe performance in all situations!

THANK YOU

Image Courtesy : Wilhelm Gunkel , Unsplash.com

Friends,

The book on inspection is fairly admired as true image of work being done ! The documentary explanation and observed images create an environment of faith and prosperity !

The images used in the book are relevant to subject under discussion and they are stock images . They are taken from websites such as 'Unsplash.com', 'Pixabay.com','Pixel.com'. The name of the image creator is written below the included image! We are grateful for this courtesy!

Thank You!

INSPIRATIONAL

Image Courtesy : Alice Donovan , Unsplash.com

INSPIRATIONAL

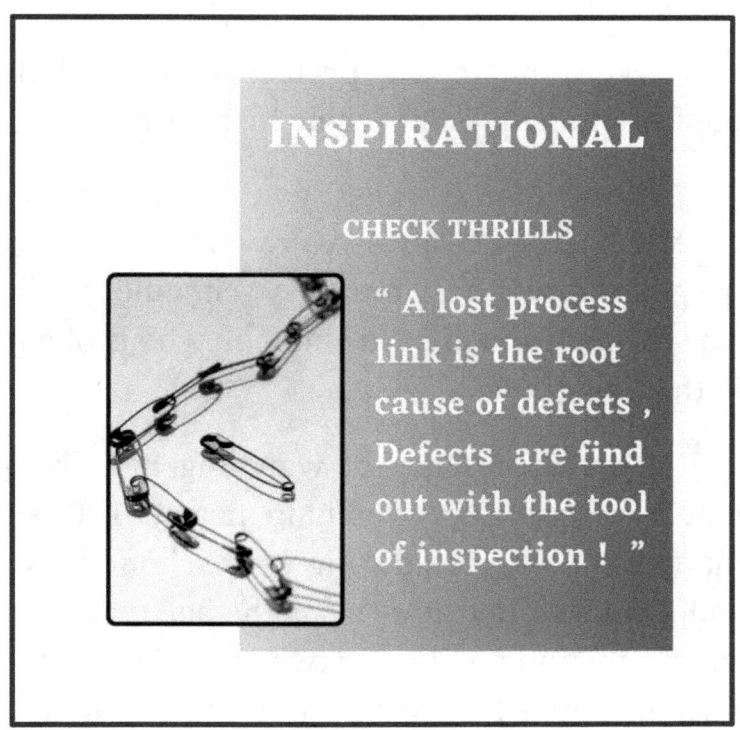

Image Courtesy : Anne Nygard, Unsplash.com

PREFACE

Hello Friends ,

Welcome to the preface section of this book written on Engineering Inspections !

How many times you watch current cricket matches ? The answer will be , whenever we get a minute from our busy schedule , we just surf through scoreboard and look who is playing in the field !

This experience of watching live cricket match comes after watching nearly 100 plus matches of all types during your school days or college days ! Because this is the age where you can easily find an hour or two for only watching cricket match ! Once you become a grown-up adult , you are fed up with more responsibilities and hence you have to devote your productive time to fulfilling those responsibilities ! Isn't it ?

Let's take another four -five examples to simplify the plot of this preface !

In this second example , you have to tell how many times you go to your vegetable market to bring the farm fresh vegetables for your daily homemade tiffin ?

This is highly personal question and people may like to answer this question once they understand the concept of inspection !

Food is the basic human life requirement and everyone wish to get highest quality vegetables and other food items at reasonable price ! This is because , the production scale of food items is mammoth and in a country like India who is bestowed with strong agricultural footprint , there is fair possibility of quality vegetable and farm product production if suitable financial and environmental back up is available ! Indian farmers mostly use traditional wisdom of producing crops and they have their peculiar style of celebrating various crop progress phases milestones ! People know these celebrations as a part of Agri friendly culture !

So, the first observation done during buying your farm fresh vegetables is you look at four five key things!

These are related to name of the product, its color identification, its overall smell in display zone, its variety as per different size and shapes and its price of buying!

There are few offers given by farmers for buying higher quantity of vegetables at discounted price, this is because, the product has limited lifetime!

Will you buy vegetables which are available at 50 % price than its original purchase price and they are not looking fresh and healthy ? Certainly No ! Not at all ! The food is considered as most nutritional element of daily life and it get formed with farm fresh quality vegetables ! It directly function or malfunction your existing body system and to keep yourself healthy and prosperous all time, it is desired that you must choose the vegetables which are best for your overall health card ! Isn't it ?

In third example , let us consider the buying medicines for your loved ones in times of their sickness ! Do you go to medical shops or clinic personally or order the medicines and clinical advice 'On Line '?

The answer of this question is in a typical period of one year , every healthy human being in typical health group between young age to adulthood , either visit clinics once or twice in a year ! There is health-conscious mass who daily take care of their food , their travel and their work-related deliverables in such a way that they remain 100 % fit throughout the year and they don't need to buy any single pill for their health-related worries !

On the other hand , there are many sick people , who doesn't bear good health aspects . Their frequency of going to clinic is regular and if they do not visit clinic as per prescribed visit call , adverse effect on their health may happen ! Such people are clinically and psychologically managed by their family physicians , regular consultants and known family members !

What you prefer to check when you buy medicines for yourself ,your dear ones or any stranger who is seeking your help to buy the prescribed medicines ? If you have ever lived in villages , you will frequently need to help such strangers this is because they are not aware of the language in which medicines are prescribed ! See , the fundamental role of communication language in the process of inspection !

You simply hand over clinical experts' prescription to medical shop pharmacist and they check the stock of medicines available in their drug inventory which is specifically arranged in dedicated rows and racks ! They access the medicines , check its expiry date and provide you bill of medicine purchase !

You personally verify all the available medicines and confirm from pharmacist about schedule of doses as prescribed by clinical experts or doctors ! You do check the quantity and quality of the medicines in the form of its expiry date ! A chemical either loses its reactivity or increases reactivity with body elements once the expiry date of medicines is over ! It depends !

Once the medicine is consumed as per suggested schedule , your body mechanism react to their inherent power and the present microorganism in your body behaves in such a way that it regains its original health status ! This is basically known as recovery and it can happen in one day or after several years depending on nature of disease faced by person or patient !

If there is no favorable effect of medicines , you again revisit the clinic and doctors prescribe other set of medicines or they enquire how you took the prescribed doses ! Isn't it ?

On revisit , you probably get the expected dose and you recover from that illness . If you are not satisfied with medicines or clinical expert , you may believe and visit regularly or you choose to look for different expert in your nearby or remote area ! Isn't It ?

'Health is wealth' and hence you do not like to take any risk with your health for its short term and long-term protection ! You preserve health and keep yourself healthy and active so that you can lead a productive life !

In fourth example about the concept of inspection , let us take an overview of your first air flight ! Is this, okay ? Maybe , everyone may not experience this type of task before but it is fairly interesting to discuss the process of air travel with regard to inspection !

You have to go from one place to another place almost within a day and you choose to go by Air way after confirming the tickets are available ! You get all the details of your trip and expected arrival and departure to two cities !

With fulfillment of your travel related required documents , the airport authorities or administration allow or disallow to on board the flight ! You are checked for your belonging and your overall fitness for travel. If you are travelling for medical reasons , then you have to follow laid down protocol to ensure there is no emergency during your travelling ! We often hear about some special experiences related to medical condition during air travel ! The chances are least but they are there ! So , once you start your journey , you are instructed by flight crew and on landing ,landing formalities are started!

In this matter of time of let's say 2 hours to 12 hours , you achieve your ultimate goal of reaching the desired destination on time and hence fulfill the task for which you choose to travel !

You experience some travel related uneasiness which is specific to travel , you also experience that during your journey , the constant instructions by flight crew are given to make your travel experience easy and enjoyable as far as possible ! You experience , there was strict verification of your on boarding passes and other documents of identity and there is luggage verification before and after your journey by dedicated officials ! In all these security checks , if you are not questioned for any unsolicited reason , it is known that , your identity is authentic and you can certainly travel to different country or within your own country !

However , if there are any irregularities regarding travel protocols are observed , you have to face enquiries from airport security officials and in case you found guilty , you have to bear the punishment prescribed in designated rules and laws ! Isn't it ?

In fifth example regarding inspection experience ,let us consider the typical home buying experience !

Home or house is one of the basic human needs and its takes lot of money to purchase your own home in leading metros ! The budget is sky high and the rooms purchased in a flat consist of four to five in numbers in a lavish and well-furnished apartment !

But how do you buy your home ? Just by referring to great advertisements shown in leading daily and websites or by personally carrying out site survey in the interested area ?

The answer to this question is home buying is a lifetime experience and advertisements act as primary source of information that gives pictorial understanding of the project and about its landmark location ! You may get basic plan of room allotments and other amenities design in those advertisements ! How many advertisements to be seen is an individual choice !

So, after watching these advertisements , you shortlist 5-10 projects and visit those projects individually as per time available with you . Here you have to take care of your requirements and your budget and also the availability of the flat within specific time limit !

You check statutory requirements of the projects and also see how many banks actually finance that project for home loan ! When home loan is available , it reflects , project meets compliance requirements and general professional guidelines applicable with the project !

Then you see the quality of construction and convenience of project from your office and other important city centers ! It is desired that project need to be near major city establishments so that there will be very very less time will be provided to visit these places regularly !

Your next task is finalizing the flat and apply for home loan and arrange for down payment !

Once the downpayment is done and your home loan is approved , all you have to do is payment of applicable EMI before possession and after possession ! More payment you do before possession always help you to reduce the later burden of payment after possession ! Then on one fine day , you receive the key of your flat and become a member of co-operating housing society with which your all-social interactions within that housing society starts !

In this whole process , you have to inspect many small things for a period of more than one year ! Some stages if not get completed on time , then you have to wait for final possession . These moments require patience and you need to bear that time period ! Secondly , once you have decided to purchase your home then for a period of 10-15 years , you have to pay the EMI of your house and hence your financial responsibilities increase ! You have to maintain a tight budget if you are purchasing other luxuries like car , jewelries and foreign trips ! So , you are not just buying a home , but you are carefully taking a call about the possible effects on your overall lifestyle !

The cost of wrong decision making in typical home buying experience is high and if you fail to live in good ,calm , helpful society , then you and your family members has to struggle for small small things ! Hence ,selection of proper area is very very important !

As you progress in your life , you may purchase another flat ! This time may be bigger and smarter ! This helps you to upgrade your lifestyle to new level and you try to match with latest living standards !

So , the learnings of this example are home buying need keen and detailed observations and one must not miss essential facilities in society in which you are going to live for rest of your life ! You need good financial acumen and you can always find the best rate for best flat in best locality !

Home means safety of your own and your family members . Compromise to home security is highly expensive and you always try to keep your family in safe and fun-loving society ! Hence , the detailed inspection is must !

In the last example to understand the concept of inspection , let us take an experience of earning a post graduate degree from reputed foreign university with provision of scholarships and job assistance !

How do you start this process of university search ? You visit respective websites of specific universities and try to arrange as much information as possible ! In today's time , universities provide lots of information about courses , staying arrangement , fee charges , facility listing and things like that ! They also specify, the typical entry criteria and entry exam needed for permanent admission !

After this information , the fee structure and other financial credibility requirement are well explained and accordingly student prepare their admission choice list !

There are two things for such type of admissions ! First is you need to clear the admission criteria and required qualifying score! Secondly you must pay the college fee to confirm your admission !

Depending upon the nature of your professional course , you have to spend those many years in the university and ensure you score well in every semester ! By the end of your all semesters you get your professional degree and also get opportunity to find a best suitable job rewarding your qualification ! Your life changes and after getting permanent job , you keep earning descent money and thus start living an altogether different lifestyle !

Here , you have to start your university search when you are studying in your 10th or 12th grade division ! Once you prepare for entry exam and appear for it on available platform , the university administration declares test result as per indicated schedule and thus provisional merit list is declared . If your rank is promising , you are called for interview process and after satisfactory answering , you are awarded your seat in that university !

The Academics and professional development are two important things in anyone's life and hence you have to carefully choose the best path of your academic interest and future career ! If this path is selected correctly , then your progress is confirmed if you maintain desired track record !

In case , you didn't get the university of your choice , you have to give preference to your career interest in other university . It's absolutely fine ! The scheme of education will be little bit different but the course deliverables are almost same !

But you should not compromise on your professional interest for a specific university ! Your education and university education are basic foundation of your overall career . So , you have to choose course of your interest and also which has good future earning possibilities !

Here , you have to check , other essential details like number of flights available to university location , the local culture and climate , food habits and other hobbies ! You have to adjust a lot to new conditions and also ensure you score well ! You have to visit your motherland on vacations and on return path ,you have to again match with typical conditions !

Now let us understand two-three examples of daily professional life where inspection is not done and things are used as they are ! This will significantly simplify the purpose of carrying out inspection !

What happens to a weighing scale when more than prescribed load is loaded on the machine ? This is quite a common experience at many weighing stations where raw material stock is weighed !

If the weighing scale is calibrated properly , the operator will immediately note the extra load by using his day-to-day experience of weighing ! Then he will ask to reduce the load and then weigh the stock ! If the overloaded vehicle is weighed , it certainly affects the performance of weighing scale and as a result of which it may indicate false values and thus there will be two different reading for same load at supplier's place and at receiver's place ! People will enter into conflicting mode and then they have to resolve the issue by weighing the same stock at a third-party weighing scale ! Here after noting final reading matter is resolved !

So, what is the takeaway from this example ? If you are not checking material correctly at weighing scale , the overloading hampers performance of weighing scale and shows conflicting readings ! Suppose a truck is carrying total weight of 5 Ton of steel plates amounting to several lakhs and if one weighing scale showing wight less by 100-200 Kg accounting to saving of few lakhs in invoice , the supplier will start arguing with purchaser and will claim that he has supplied 5T material and he will not accept a single rupee less than the billed invoice ! On the other hand , purchaser has to agree the supplier's statement is correct and there is defect in his weighing scale which is later resolved and the material is weighed again after repair and the weight is found within allowable tolerance !

This is one example of not carrying out proper inspection at weighing scale station . Before weighing load , operators have to ensure the weighing system is calibrated , dummy weight of sample weights are tested before the actual testing and the difference in weights is noted and it is within allowable range ! After such type of trial calibration , they have to go for real weight checking !

In second example , suppose you have supplied 100 columns to a work site nearly 2000 kilometers away from your manufacturing place . The work site has also requested to supply 1 extra column per 10 columns for any adjustment related purpose ! While sending columns , this requirement is not noted and only 100 columns are dispatched ! The required quantity was 110 while supplied quantity was 100 !

When the actual site fitment started , it went smoothly up to 90 columns . In last 10 column , for finishing stage adjustment , 3 columns were in need of extra support of three columns ! Noting this scene , the work gets halted and people started seeing how this is short supplied when the instruction of extra column was quite clear in the work order ! The site engineer and dispatch engineer started talking with each other about this short supply and its reason of miss out ! The dispatch engineer agreed that same is overlooked and extra quantity is not reflected in packing slip as same was not part of engineering bill of material!

When they reviewed the engineering bill of material, they observed the footnote where it is indicated that after every 10 columns, you have to supply additional 1 column for site adjustment ! Now when the packing slip is prepared, only tabulated quantity got included and this extra requirement is missed, as a result of which, the material is short supplied ! However, the person who manufactured 100 columns noted this requirement and he has prepared 110 columns and dispatched completely ! When the material is unloaded, the operator counted 100 columns as per packing slip requirement and keep extra 10 columns at that place only !

After discussion with site engineer, dispatch engineer inspected that area and he found 10 columns lying in dispatch zone ! He picked up three extra columns and sent it to worksite 2000 Km away for place of manufacturing ! The material reached site after one week time and installation completed in next two days !

(XXX)

So , what is the takeaway from this example ? The answer is , you have to thoroughly study the drawings and accordingly count your bill of material ! Footnotes are specific and you have to amend your packing slip once you need certain additional material ! You have to note that in case you short supply material to work site , the work gets delayed till the supply is fulfilled ! So , take care to comply with dispatch scope !

In another example , suppose two identical jobs of same design are getting dispatched at two different places . The jobs are identical but few interconnecting components are customized and they are designed for two different sites !

The site 1 has interconnecting pipe measuring 2 meters while the site 2 has interconnecting pipe 2.1 meters ! The site 1 is located in south part of the nation while site 2 is located in north part of the nation ! Now what happens , because of identical jobs , the most of the supply is same except the interconnecting pipes ! During packing , by mistake , the interconnecting pipe measuring 2 meters get

Packed in supply that actually requires 2.1-meter pipe and 2.1-meter pipe get packed in job that requires 2-meter interconnecting pipe !

Both jobs get dispatched on same day within 15 minutes time gap and one goes to south direction while other goes at north direction !

When their site assembly is started , the operator note he has received different interconnecting pipe and it is measuring 100 mm extra ! He calls dispatch engineer and ask for this wrong supply !

Dispatch engineer checks his packing slip and part number and he notes the material is wrongly interchanged to other similar job ! He conveys the message to site engineer that there were two identical jobs dispatched on same day and somehow the pipe got interchanged ! He immediately calls the other site engineer and ask to verify the length of pipe ! They measure the length and communicate the dimensional error ! Now , where there is excess pipe available , it is cut to desired dimension and fitted to assembly

Where there is short pipe available for assembly , 100 mm extra pipe is procured from local market and it is joined with original pipe with a groove weld ! The joints radiography and dye penetrant test and hydraulic test is done and it is connected to main assembly !

Since it was easily available pipe , local market could provide the same ! If that pipe was special , then local market may not have supplied it and you have to arrange it from different market !

So , what is the take away of this example ? The answer is , identical job can have nonstandard changes and one has to measure every dimension , their part numbers before sending the material to site !

Another take away is even though there are trouble at site , same can be managed through local market supply and if the material is not available in local market, then you have to wait till the correct material is arranged by responsible engineer ! The cost of not doing correct inspection is thus provided advantage to local market !

In another example , you have to measure nearly 1000 loose pipes for their actual length to decide the maximum length of required pipe to pipe joint ! The total required pipe joint length is 4 meters and the pipe available has length varying in between 1-2-3 meters ! You have to measure each pipe and note its dimensions and final length ! After this measurement , you have to join pipe with each other in such a way that there will be minimum pipe to pipe joint and least wastage of pipe material after cutting for required length !

Now see the extent of inspection needed to be done in this example ! You have to first measure all pipes and segregate themselves in variety of 1-2-3 meters ! The joint pipe length is 4 meters and this length can be achieved by joint combination of 2+2 , 3+1, 1+1+1+1, 2+1+1 ! As you have to ensure minimum wastage of pipe during precise cutting and you also have to avoid excess joints ,so the combination of 2+2 , 3+1 are most preferred lengths , once these lengths are finished in total 1000 pipes , you have to go for

Another combination that requires two joints, which are 2+1+1 ! In the last you have to think for the option of having 3 joints with pipe length combination 1+1+1+1 !

You may be aware that welding a pipe to pipe joint with the help of radiographic quality joint welder and getting it tested by radiography requires huge investment of time and money and hence if the total joints will be less, the joint testing expenses will be lower !

Now, what will happen, if person not segregated the pipes and started welding them just to form pipe joint of 4 meters ? There will be more joints and there won't be any specific combination . This will make wastage of pipe because of sized cutting and there will be increased number of joints ! Because of this, the cost of testing and welding will increase and you have to also think for radiography repairs if any ! The cost of not doing inspection is so high that it is always recommended to carry out basic raw material inspection before starting production of any project activity ! So, what is the takeaway ? Engineering combinations help to reduce

Wastages and they keep number of joint in minimum required zone !

' Check Thrills ' is about learning all such types of practical experiences in simple and interactive language ! When you describe the incidences matching with real life , the understanding become clearer and more certain !

Throughout this book , we have tried to portray the need of different types of inspections , different tests , different observations just to create an excellent quality safest job ! When quality and safety are available , then you don't have to think for any performance related issue !

Check thrills will also provide you standard guidelines for different types of inspections so that in case you are not aware of these tricks , you will come to know their importance in the field on Engineering Inspections ! Let's move ahead with different topics of this book which display the knowledge , skill and experience required to carry out any inspection ! Thanks!

INDEX

Sr.No.	Description	Page No.
1	Thrilling Height	1-10
2	Thrilling Depth	11-20
3	Thrilling Size	21-30
4	Thrilling temperature	31-40
5	Thrilling Surrounding	41-50
6	Thrilling Inspectors	51-60
7	Thrilling Customers	61-70
8	Thrilling Auditors	71-80
9	Thrilling Defects	81-90
10	Thrilling Reworks	91-100
11	Thrilling Approvals	101-110
12	Thrilling Sites	111-120
13	Thrilling Changes	121-130
14	Thrilling Receipts	131-140
15	Thrilling Dispatches	141-150

INDEX

Sr.No.	Description	Page No.
16	Thrilling Calls	151-160
17	Thrilling losses	161-170
18	Thrilling Heat treatments	171-180
19	Thrilling Painting	181-190
20	Thrilling Packing	191-200
21	Thrilling Dimensioning	201-210
22	Thrilling testing	211-220
23	Thrilling Design Review	221-230
24	Thrilling Order review	231-240
25	Thrilling On Site concerns	241-250
26	Thrilling New Project	251-260
27	Thrilling Machines	261-270
28	Thrilling Public Comments	271-280
29	Check Thrill !	281-290
30	Mock Drill & MCQ's !	291-356

LET'S CHECK !

(A-X)

THRILL 1: THRILLING HEIGHT

Image Courtesy: Power Line , Pixabay.com

1.1 Introduction:

Hello Friends,
Let's start the journey of this new book '*A Precise Commentary on thrilling engineering inspections* – **Check Thrills** – *Learnings of Engineering Inspection Services*."

Engineering Inspection is a critical process step in which decision of acceptance, rework or rejection is made ! Its completely intellectual job made with the help of basic job knowledge, various relevant references, drawings and standards, understanding of tolerances and measuring equipment's, experience of service environment and skill of operators and overall safety concern of the part being inspected !

There are various standard situations in which inspection is carried out. It can be receipt inspection, in process inspection, shop inspection, site inspection, dispatch inspection, lab-based inspection, on shore and off shore inspection, robotic inspection, etc. !

In every type, there are again sub types, e.g. in case of lab inspection, there can be 'n' number of labs such as destructive testing lab, radiography testing lab, ultrasonic testing lab, dye penetrant testing lab, magnetic particle testing lab, welding lab, chemical lab, physical lab, biology lab, electronic lab, computer lab, electric lab ! In all such places, parts are examined for their fitness for use by applicable testing and performance evaluation methods !

Same is the thing for in process inspection ! In process inspections are carried out to ascertain the product quality when product is under real time manufacturing . This helps any system to arrest any live defects on line and correct them before they settle in product being manufactured . What does it mean ?

Let us consider an example of casting industries where every casting requires metal pouring in molten condition in its allocated mold ! Before pouring hot molten metal into its mold, it needs to be extremely clean and free of any slag formed during melting reaction . If this slag is totally removed before pouring, the resultant casting become homogeneous and there are very little chances of internal defects which are usually noted during radiographic testing .

Moreover, you may say ,you get a sound casting that has a peculiar sound when you hit with mallet or tiny hammer ! But when the molten metal is not so clean and it contains the traces of slag and other metallic impurities , these particles get entrapped in molten metal and during solidification casting defects are observed which need to be removed by again starting rework on casting as per specified rework method . This consumes important productive time and also increase throughput time which affect productivity of plant ! So , a detailed eye during in process inspection is must ! It helps to build product quality safely !

With this introduction , throughout this book , let us discuss ,different types of carrying out inspection in industrial set up which help to deliver conforming and safe products for their business to business or retail customers or consumers !

In this list , we are starting the discussion with inspection at height ! What type of physical attributes are required to carry out inspection at height ? What type of mental set up is required to carry out inspection at height ? What are risks and what are general challenges of inspection carried out at heights ?

1.2 Physical attributes of inspector :

When you have assigned an inspection call of carrying out inspection at height, the physical attributes of inspector need to be strong and very strong ! A well-built body with excellent eyesight is first need of inspector's physical profile ! Inspector should able to climb ladders and walk on narrow platforms situated at height of several meters ! In this movement, he must feel fearless and must able to walk with total confidence to carry out required inspection !

There is simple technique of carrying out inspection at height when you are doing it first time. You have to first master the art of carrying out inspection on ground before you start carrying out inspection at height. This is because , you need to be aware about basic body movement such as bending , moving , fast walking , slow walking , low stretching , low jumping which need to be done when you are carrying out an inspection in the real shop floor or work site ! When you attain required flexibility , you can climb up the ladder and start moving on platforms situated upper side !

At suitable distance , your job will be located and you have to observe the respective part without looking below . When you are inspecting at height , there are chances of height phobia and because of which some inspectors initially feel little bit uneasiness ! But when you just concentrate on your job which is rested in front of you , things get better and with two-three repeat attempt , you get awareness about how to approach the job in safe and comfortable way !

One of the most important characteristics of work at height is it requires excellent grip and for this your fingers need to be dry and thin ! When you are climbing up the ladder , there are chances that your fingers may get wet and because of which ,you may find some sort of loose grip ! To avoid this situation , you are always provided with suitable hand gloves with the help of which you can hold your grip around climbing ladder bars and make your way easily !There are safety net , cage and guiding rails where you can help yourself to take stable stand

and just look around for a minute or two to settle in the environment !

1.3 Mental Frame of Mind :

When you are climbing up the ladder where resting platforms are situated after several meters , before climbing , you have to plan your upward and downward journey ! How ?

First , you have to see general arrangement and terminal point drawing of that project and see how the assembly is designed and from which point you have to approach other point !

Once you see resting platform and required steps on ladder , you have to start climbing ladder by following left leg- right leg combination ! Always ensure in drawing , is there any pitch variation for steps ? If there is ladder step pitch variation ,then you have to walk cautiously to avoid mis fall on different step pitch ! This is the most common reason of people fallout from height ! They simply start climbing and at variable pitch their foot forget right stepping and this create momentary mental disturbance and they may lose their balance suddenly and fall down ! Hence , before climbing upward or coming downward , look at for any pitch variation of steps ! Never change sequence of walking . Left leg followed by right leg followed by left leg should continue. When you have to pause in between , don't look behind , look forward , hold the bars in both hands and stop on any step for 5-10 seconds and then move ahead ! Your hand and foot movement must occur in synchronization ! You have to take care that not your hand nor your foot must slip when you are climbing up or down ! This precaution will fulfill half of the access requirement !

So , break the total steps in comfortable stops or phases and complete each phase safely to reach the destination situated at several meters !

Once you reach intermediate or top platform , take a small breathing break and feel the scenery around ,especially for height above 30 meters , the natural landscape of nearby area looks beautiful and satisfy eyes ,it also releases burden of climbing up ! Platforms are fairly wide where man-movement and portable machine movement can be done with caution ! Access the inspection point and carry out the necessary inspection with the help of required documents and inspection tools ! Record your observations , take photographs if allowed and communicate the status of inspection with your reporting authority . If they have any query , try to resolve when you are present on top of the platform ! This is good practice to communicate the status when you are inspecting nearly 40-50 meters above ground level ! If something remain balance , the reporting authority may guide or may ask ! But if you came down and if that observation is missed , then you have to wait till your next climbing ! Hence always communicate the status from top to your reporting authority or collect complete information required to make correct decision without any doubt or error !

1.4   Risks of Inspection at Height :

There is risk of nausea , omitting , height phobia , breathing discomfort , risk of falling , risk of forgetting important document or measuring equipment on top , if wind pressure during climbing up or down is high just like noted in establishments situated in hilly or sea-shore area , there is risk of falling due to wind pressure ! When you are constantly engaged on top portion , there may by sun stroke and heat waves impact ! There may be eye irritation and ear choking instances ! Hence you need to wear all PPE before climbing up !

1.5 General Challenges of Inspection at height :

Once you learn to climb up and down safely with carrying your PPE and other tool kit along with you or with the help of assisted device like crane or lift , the working method is almost same which is available on ground ! Only thing you have to keep in mind is you are working against gravity and if you fall down , there will be major injury and hence you have to limit your movements and take every step with total caution and as per set inspection protocol ! Once you finish inspection , again you have to reorganize yourself and climb down slowly!⊛

THRILL 2 : THRILLING DEPTH

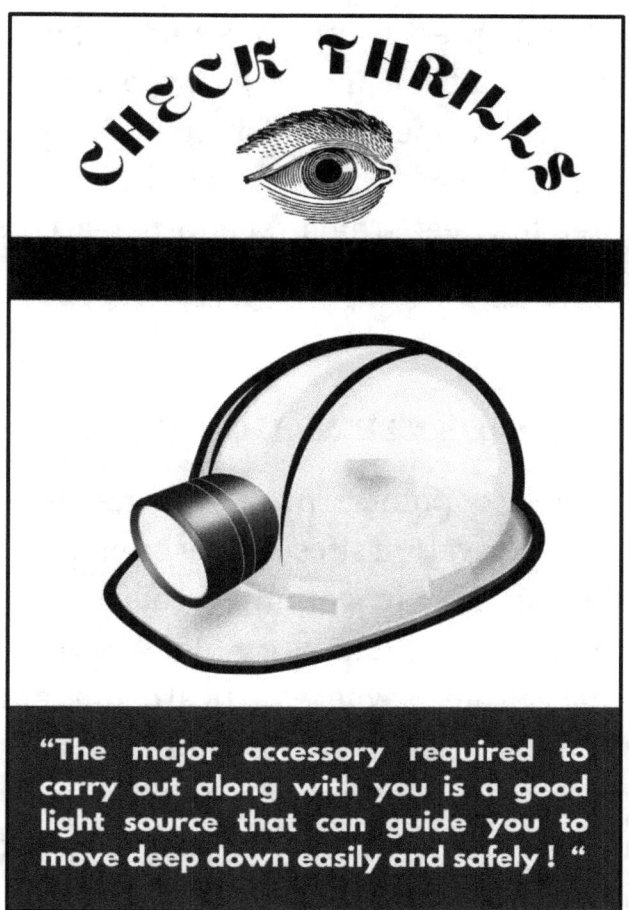

Image Courtesy: Hard Hat , Pixabay.com

2.1 Introduction :

In last chapter , we have seen , how the inspection at height is to be done and what kind of thrill is present in accessing inspection points ! In this chapter , we are spreading lights on how to carry out engineering inspections where location of inspection is situated far below the ground level or the internal confined area is located too deep from first entry or access point of the job under inspection !

2.2 Typical deep inspection points :

These points consist of metal- mineral mines , excavation sites , underground railways , underground irrigation projects , deep foundations of multi-story high rise buildings , refinery projects where vertical drums and other peripheral devices to be inspected , underground electric and telephonic wiring , underground secret path construction related to military purpose and things like that !

We have covered only deep inspection points located on natural soil ! If deep waters are concerned , there are special inspection requirements which are carried out in inspection of oil rigs and sub marine territory !

2.3 Basic Inspection accessories :

Friends , you are moving deep down the ground and hence you need to be extremely alert before carrying out such type of inspection ! There is very very special site safety protocol laid down for deep place inspection and until the responsible head do not permit you for inspection , you cannot go there on your own !

There can be any activity going on in that place which you may not be aware ! Also , you cannot easily see it beyond few meters and hence internal clearance from site safety authority is must before going ahead for deep place inspection !

The major accessory required to carry out along with you is a good light source that can guide you to move deep down easily and safely ! Also , you must be medically fit so that you should not feel breathlessness when you are moving deep down. You must handle designated communication device provided to you to carry out necessary personal and site safety related communication. In some places, there is typical way of passing communication from one point to other till it reaches top or bottom most end, you have to follow that protocol and signs of communication with each other ! In case of any trouble or unexpected situations, you are trained before going deep down about expected response from your side. There are specific arrangement points where one can assemble in emergency but every inspection place may not possess such place for implementation. So, there will be other method to tackle the apparent risk which you need to know or site people will train you in that aspect also. You have to remember those urgent tools in case of any unexpected event during deep place inspection !

Considering the effect of less light and possible sweating, some type of artificial cooling may be provided or clothes will be designed to suit deep place inspection. You have to wear that protective uniform before moving deep ! Also, protective glasses, ear plugs, hand gloves, small mirror is good catch to take with you !

2.4 Physical Attributes of Inspector :

Deep place inspection needs people with moderate to short height and normal physical strength ! Heavy weight champions and tree taller personalities will not be a recommended choice ! The reason behind this limitation is deep place inspection has very narrow access through which inspector has to pass and carry out their job . If they are heavy weight , they may not enter through initial narrow access and may stuck in the entry itself ! If the person has good height , he may also get stuck in between limited height available inside deep place inspection points !

The inspector's physical built need to be normal – not too weak , not too overweight , this is because , they need to go up and down frequently and this may put sufficient pressure on the path through which they are walking ! As these access points are arranged deep down the ground , extreme care is taken to ensure there is no deep land slide causalities and even it might happen , still sufficient alarms can be raised to control room !

Inspectors must able to walk down with the help of light source , cooling system and ladder steps which can be wide or narrow depending upon the place sufficiency ! Secondly , the pitch variation of ladder is also a point of concern . One has to see , how the steps are designed and they have to follow left-right-left-right sequence ,while going down and coming up ! For this frequent movement , inspector need to be sharp , accurate and flexible !

2.5 Mental Set up of Inspectors :

Deep place inspection is a skilled job and only professionals with sufficient on -field experience are allowed to touch this work ! No new comer or trainee engineer will be allowed to access this place unless he achieves sufficient practice & proficiency in dedicated training center . Some training centers may take written exam about the basic inspection and health hazard knowledge from every student and if they pass that exam , then they are considered as fit for that role , else they are disqualified ! This discipline is a mandatory discipline to prevent unskilled person enter the potentially risky place of inspection ! Mental set up of inspector need to be relaxed , confident and fearless !

Inspector need to be very much clear about the basic fact that he is approaching a potentially risky area and it is his passion and quest for excellence which inspire him to do this job with daily motivation and faith in his capabilities !

In typical span of 100 days , 80 days will be normal , 10 days will be demanding , five days will be challenging and five days will be extremely difficult as far as field conditions are concerned . Accordingly , inspector has to plan their work schedule so that safe inspection can be carried out ! In no instance , his personal safety should be compromised from his own side , he needs to be sufficiently careful & alert !

In seasonal variations , necessary alerts are raised by field administrators which they have to adhere and plan their work accordingly ! Site work always has limitation of site conditions and hence it cannot be fixed in definite target such as work carried in enclosed shop floor ! Hence inspector has to adhere to site safety protocol before entering into deep down area !

Inspectors has to raise his concerns in case necessary preventive and corrective maintenance is not done by the field authorities . In deep down work sites , maintenance act as necessary site lubrication that allows free movement of people and resources from top to bottom and bottom to top !

2.6 Risks observed in deep down inspection sites :

As we know , deep down inspection is comparatively riskier than inspection done at height ! There is risk of fall, in inspection at height ,but if someone stuck somewhere at height , he can be brought down by sending suitable aerial solution , but if somebody stuck deep down the site , how one can see , he is stuck in the site unless his sound reaches the site surface ! Hence , proper alarming and interlinking with control room and war footing stage is extremely important !

There is always risk of non-functioning of light sources which create dark environment and which make work extremely challenging . Inspector must able to withstand this delay till sufficient illumination is received ! Light source need to preserved till the completion of inspection !

The other risk consist of is risk of short of breathing or lack of sufficient oxygen supply either natural or artificial ! Many incidences are recorded where short of breathing become cause of extreme causalities ! Hence illumination and oxygen supply are two important life savers in deep down place inspections !

Presence of poisonous reaction gasses and their sudden release in working area is also considered as major risk for inspection and working ! There are chances of land slide or component part fall out because of excessive corrosion and weakening ! Every step has to be taken carefully !

In some instances , people may go in wrong direction and may stuck in the middle , in such unfortunate events , every effort need to be done to rescue such people by all possible means ! Recently one of such incidences happened in India during tunnel preparation in which workmen stuck inside and some portion of mountain fall down which obstructed regular entrance of the deep work place !

2.7 Challenges during inspection :

It is extremely difficult to provide the decision of acceptance or rejection based on observations done in deep down inspection , however every care is taken to complete most of the job when the parts are accessible on surface ! Designers try to design parts in such a way that inspection area should be sufficiently accessible and ventilated .

Manholes , Head holes , Inspection Covers , Access doors , front doors , front and rear chambers are provided to carry out necessary internal deep-down inspection ! Deep down parts are inspected during manufacturing by placing them horizontally on rotator and through manhole inspector inspect the internal completeness along with a person staying outside for inspectors' safety , light source arrangement and ventilation supply monitoring ! Big fans or blowers are used to provide ventilation air to workmen and inspectors working in confined areas !

One thing satisfies this hard work on deep down inspection – Everything is complete in all respect internally & deeply ! ✪✪✪

THRILL 3: THRILLING SIZE

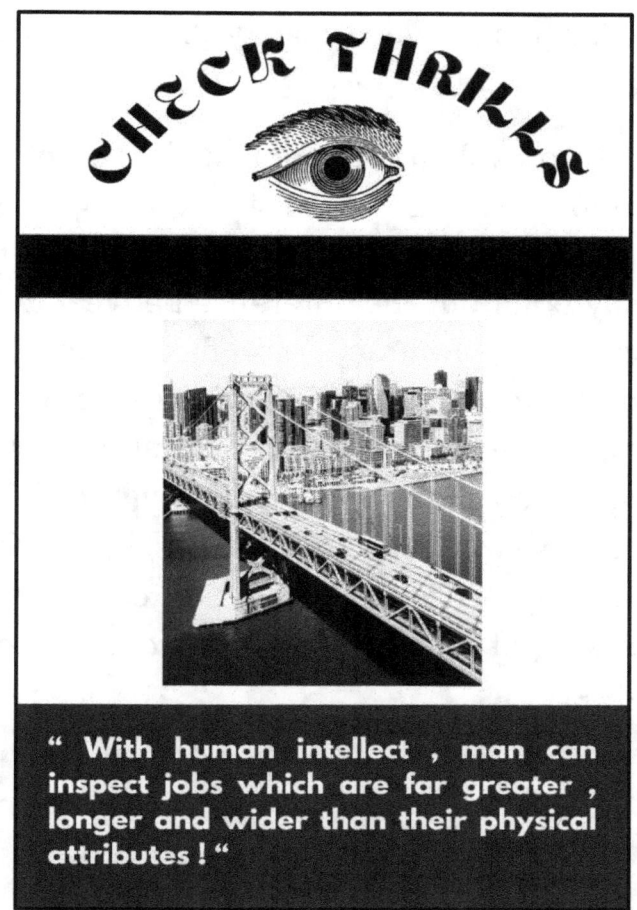

Image Courtesy: Denys Nevozhai , Unsplash.com

3.1 Introduction :

Size matters everywhere and size is the indication of human efforts put in the design of product or service ! At the same time , size refers to higher or lower capacity of that product or service !

To explain this point with the help of an example we can consider design of an automobile and design of an aero plane ! One can easily point out the difference between product size , service accuracy , risk of usage , price of the product and passenger carrying capacity ! Hence , inspection tactics required for automobile and that for aero plane are comparatively different as their life hazard class is different ! In this chapter , we are discussing , inspection carried out for checking of mega size and micro size products and services !

3.2 Physical Attributes of Inspector :

When you are carrying out inspection of mega products whose linear and circumferential dimensions are in several meters , inspector need to be strong and very strong !

When inspectors stand in front of such mega structures , they simply look as a toy standing in front of a big building ! However , this is what the magic of brain power is ! With human intellect , man can inspect jobs which are far greater , longer and wider than their physical attributes !

On the other hand , when you are inspecting extremely tiny products which are not visible by naked eye , inspector need to be sharp visioned ! Here these products look as if they are toy standing in front of mammoth man !

So , though it is mega structure inspection or micro structure inspection , inspector need to be strong observer and prompt reporter of whatever he or she observes during inspection !

3.3 Mental Frame during mega and micro size inspection :

During mega part inspection , inspector has to think that this mega structure is made up of certain number of small building blocks and when all building blocks are ready and accurate , then the structure is assembled by joining small building blocks ! So , if I am able to inspect each block correctly and, in the end, the complete assembly wholistically, it implies , the mega structure is accurate as far as inspection is concerned ! So , the fear of size vanishes and it get converted into clear cut logic of inspection of individual building block !

With this logic , every mega structure is considered as a system which is consisted of other sub-systems ! Every sub system has typical assemblies and these assemblies are interconnected with each other with the help of components , channels , pipelines , control stations , discharge points , input point processors and such technical arrangement which deliver a particular process task in that sub system . So , the functioning of main system depends upon functioning of other sub systems either jointly or independently !

When you are inspecting micro structures and micro dimensioning , you have to keep in mind that you are handling high precision work where allowable tolerances are extremely and extremely narrow . You have very few chances of slippage and you have to hold your decision tight if you observe any deviation beyond allowable range!

The fundamental technique used in the inspection of micro size part is use of largest scale available with microscope or inspection device such as digital screen ! Since, the natural size of part is so small, therefore people convert it into visible size with the help of microscope and Lense technology ! In this inspection, the whole idea is to show the part with more fine details so that you can understand the interlinking between various system elements and hence you could derive the necessary inferences of that inspection !

In micro part inspection, you have to keep in mind the scale of inspection ! The defect if any is far much shorter than it is visible in microscopic observation ! Hence, once you finish this inspection or observation, you must co-relate the size of defect to true scale so that you can guess its overall impact on original size of the product !

In clinical microscopic inspection, clinical experts evaluate the size of natural organism and accordingly plan their counter solutions ! May it by microscopic inspection , ultrasonic inspection , radiographic inspection , all they need to see is actual size of particular foreign entity which is causing harm to human body structure !

3.4 Risks observed when inspection is carried out with macro sized and micro sized parts :

When you are carrying out inspection of macro sized parts , you have to keep one thing in mind that you need a constant companion along with you to help yourself during inspection ! Macro sized parts has huge dimensions and you alone cannot inspect these dimensions and structural features ! The best technique applied is a pair of inspectors undertake the inspection and communicate with each other before making observations of that inspection . The responsibility of that inspection belongs to both and no one can deny any adverse remark noted after inspection ! For this reason , they have to jointly communicate the observations and record them with proper verification ! So , first risk is , you can't do complete inspection alone , you need help for at least one dimension !

Second risk is you need proper supporting tables and inspection bays specifically designed for inspection . The macro job will be arranged on such dedicated inspection bay and from there you have to carry out particular stage inspection ! It means , if you have to take radiography , the job will be moved to closed enclosure and their test will be carried out ! If you have to test water leakages , then job will be moved to hydraulic testing bay and there you have to inspect job for any leakage when test pressure is applied . When you have to inspect job for painting , you have to shift that job to painting booth and when painting is over , it will be shifted to nearby painting inspection booth ! So , the risk is , you need to aware about the basic safety requirements of all such inspection areas and you have to adhere to those safety norms before entering into that inspection bay !

Third risk is errors in inspection ! For macro sized jobs , you have to use longer tapes and longer plumbs ! There are chances that as the length goes on increasing , there might be error in recording actual dimension which create issues after complete assembly ! Even though a dimensional variation of 20mm for 20-

meter job create major rework if it is noted at crucial mating point ! Carrying out rework on such mega structure is a long duration work ! Many people fear to inspect huge jobs because of this risk ! The jobs are expensive and even though minor errors happens and if it requires major rework then the cost of rework eat up your profits even before you manufacture your product ! Hence , you have to be very very cautious when you are inspecting macro sized products !

In micro sized product , the first risk is absence of required phase in microstructure . Microstructures are inspected for detecting any abnormality in material internal structure and their standard dimensions ! If the standard phases and their dimension is not observed then you have to find out the distorted version and the amount of distortion ! You have to find out possible process cause of that distortion and accordingly you have to suggest necessary remedy to regain the original microstructure !

Heating cycle design and cooling cycle design along with addition of specific catalyst , inhibitors , reactants are done to deal with challenges of micro structure related issues .

Another risk observed is noting the accurate dimension and its allowable tolerance ! The micro dimensions are in micron and they are specifically intended for typical surface finish requirement which is special characteristic of that product . Hence , those dimensions never fall beyond given tolerance limit . If they fall , the parts are rejected !

Micro sized parts are used for very very special function and they can be fitted in human body for clinical procedures . Hence , not only inspection but overall part production has to be done according to set clinical production protocol . So , in case of deviation , there is big risk of human life loss ! The defective part will not perform its intended function and it will risk the system in which it is installed !

3.5 Challenges during inspection of macro and micro sized products :

Challenges associated with macro sized products includes availability of site or shop conditions ! Mega structures are occupied in huge shop floors where only typical mega structures are manufactured . These shops are special and they require typical safety protocol to be followed by every authorized entity-entry !

Frequent part movement is regular affair in such shop floor and sites and inspector need to be alert of these movement . They must keep themselves safe when mammoth parts are moved from one place to other . During climbing ladder , during noting dimensions on height , they have to take care of right balance and right plumb , so that there will not be any error !

Challenges noted during micro structure parts are availability of typical code and standard dimensional profiles . Micro structure requires phase diagrams to understand the number of phases observed in typical material at various temperature and time combination . Then you can co-relate observed versus standard micro structure . If the standard structure doesn't match to observed micro structure , it means , there is major deviation in material supplied and you have to collect all proofs regarding the method of manufacturing !

THRILL 4: THRILLING TEMPERATURE

Image Courtesy: Foundry , Pixabay.com

4.1 Introduction :

When you understand a particular industry structure , you notice presence of activities those are directly related to application of heat energy for various types of internal material conversions before finish products are produced ! So , as far as heat energy is concerned , industry works in pyrometric range of temperature which are hundred to several thousand degrees Celsius depending upon the type of product manufactured . On the other hand , the industries which directly deals with freezing zone or deep-freezing applications , they also handle subzero temperature range starting from minus four to minus two hundred and more ! Carrying out inspection of parts under manufacturing of these extreme temperature zone needs special training , special PPE and special skill ! Let's glance through the check thrill of inspection with range of temperatures !

4.2 Physical Attributes of Inspector :

Inspector engaged in hot or cold zone temperature range need to be strong and fearless as well as extremely alert about surrounding temperature ! Personal protective equipment's carry the responsibility of keeping inspector safe in such working conditions and hence inspector must use PPE without any bias !

There are two-three things that need to be done before approaching the exact hot or cold zone ! First , you have to wear the uniform recommended for hot or cold zone inspection . Then if you are going for in process inspection such as witnessing metal pouring , witnessing mixing of elements as per calculated weight , extracting test pieces from molten metal , you need to notify the working team about exact time of inspection . So that they will do safe arrangement of this inspection . In the allocated area , you have to stand and carry out your observation .

Second thing, once you complete required inspection , you don't have to stay there and ask doubts to people around ! No , this has to be sincerely avoided ! Whatever your observations are , you have to note , remember in mind and write down in your note pad after coming little far from hot or cold zone ! Then in normal visiting area , you can discuss those points and seek clarification !

If you are carrying out inspection of metal pouring on a rotating mold , you have to take care that there is safe handling of metals from furnace to mold and while pouring people are using PPE ! Metal pouring is hardly one minute or two-minute work because metal start solidifying once it come into contact with normal temperature in the vicinity . Hence in this short time , you have to be extremely careful about the use of PPE and required temperature of molten metal !

If you are carrying out inspection in cold zone , first thing required is use of weather friendly clothe which are warm and easy fitting . You have to cover your head and ear safely so that effect of cold range will be least . You have to wear hand gloves and good quality long length shoes which can cover most part of your legs ! You have to maintain your body temperature normal by taking hot drinks , this will also avoid shivering and shaking of your hands and legs !

When you are carrying out inspection in cold zone , you have to ensure , you are carrying out sufficient capacity light energy source along with you or you are choosing time of inspection in such a way that favorable natural light intensity is easily available till your inspection get complete ! These are typical weather conditions and in extreme weather events , there could be snow fall, heavy rain and things like that if you are engaged into any type of site inspection located in cold region !

4.3 Mental Frame of Inspector :

As we have already discussed , inspection is comparatively risky job and inspector needs to be mentally tough and alert every time ! While carrying out inspection in hot or cold zone , inspector need to be aware about human body anatomy and need of frequent self-initiated rest breaks if he feels uneasiness because of prolonged exposure to high or low temperatures ! As per clinical advises , various industrial facilities have to arrange necessary safety assurance plan in case of temperature related field causalities are observed !

Suppose , you & your two colleagues are deputed in a steel mill where on day-to-day basis nearly 20 metal pouring operations happens through five dedicated furnaces ! For every pouring , you need to witness that pouring as a quality control engineer to see the set parameters are correct and charge is correctly melted . You also have to witness the spectroscopic record of the particular pouring to check the required grade composition is satisfactorily achieved ! In case any abnormal deviation is noted , you have to immediately carry out the correct pouring and keep the wrongly poured metal into rejected bay for its further rework and remelting corrections ! Its online work and you have to remain in close proximity of furnace environment !

In typical hot furnace environment , you may note , there are instances of metal splashing outside the furnace cubicle , there are instances of metal fumes going upwards Fastly , there are instances of metal fall out on ground during its transfer , there are instances of minor metal shocks in which slow-fast metal movement create little turbulence in mold which create risk of metal falling out , in case you are witnessing any heat treatment followed by quenching , you can see hot red mettle is quenched in cold water or resin and how the sound of metal part changes when metal is quenched and how surrounding is protected from water fall out outside the restricted zone !

For ease of inspection , all the applicable controls and their indicators are placed at safe distance from where monitoring can be done . LED indicators with suitable size are placed near internal thermocouple from where temperature can be easily seen ! Inspector need to be five to ten feet apart from such pouring or quenching operation and they have to just ensure operators are carrying out work safely and within prescribed time ! Fast pouring than required and slow pouring than required , both are undesirable conditions and hence the best time of pouring as per industrial practice is to be followed !

In a particular shop there can be combination of hot and cold zone inspection and according to inspection priority list , inspector has to visit that area with proper PPE ! One has to note the effect on human body because of exposure to two extreme temperature zones ! People are aware about how they feel when they move inside an air-conditioned room just after visiting furnace station and they are also aware about how they feel when they move from AC room to furnace station ! This temperature gap needs to be sustained by body and for this inspector must exercise on routine basis to improve their natural immunity power ! It's must ! Inspector with weak immunity could not handle thermal or subzero cyclic changes and may fall sick within few days of service !

4.4 Risks observed while carrying out inspection in Hot or Cold zone :

There are different types of risk in case of inspection at two different zone ! When you are carrying out inspection in hot temperature zone , you start sweating and dehydrating . Hence there need to be sufficient cooling arrangement for inspectors as well as you must keep yourself hydrated properly ! The average time spent at inspection site should not exceed ten to twenty minutes , if it exceeds , you must take a short break so that your body adjust the rise in temperature ! Prolonged inspection in hot area affects your body system adversely and you need regular clinical check up to see the functioning of critical body organs !

Apart from this , thermal or electrical shock and chances of metallic explosion always lye there and because of which every process step has to be ensured correctly and sufficient firefighting training to be given to every person dealing directly with hot metal or hot machines !

Provision of fire alarm , sand buckets , water pump , fire extinguishers are extremely essential and people must be trained regularly with the help of on field test trials to deal with such emergencies . This help to handle real time situation tactically !

When inspecting in deep freezing zone , there is risk of adverse effect on health ! You need to be properly ventilated and body warmers must be preset there till your inspection get complete ! There has to be specific arrangement for checking your blood pressures and heart beats to ensure your heart is working fine in cold zone ! If you are hammering something for test , ensure your fingers are at safe distance because hammer blow in cold zone hit very hard later ! You have to keep yourself safe from any type of blood clotting instances !

4.5 Challenges during hot and cold zone inspections :

When you are carrying out inspection in hot or cold zone , you have to first adjust your body to conditions and then record your observations which can be acceptable or rejectable ! If you can adjust swiftly , then the rest of the thing is just an adherence to laid down discipline !

You should not skip any error during hot or cold working . Because once that instance misses and if required action doesn't happen , then you have to again start from initial step , there is no intermediate steps for adjustment ! Once metal poured , you cannot add micro constituent later , you have to ensure your accurate ladle chemistry just before pouring ! Same thing for deep freezing zone ! If the temperature increases beyond deep freezing range , part melts and their shape changes !

THRILL 5 : THRILLING SURROUNDING

Image Courtesy: Ability , Pixabay.com

5.1 Introduction :

How does it feel , when you are sitting in a multiplex watching your favorite weekend movie ? Cool and relax ! Isn't it ?

How does it feel , when you are enjoying your evening bath in your favorite swimming pool ? Cool & chillax ! Isn't it ?

How does it feel , when you are climbing on top of your favorite natural hill station ? Cool & relax ! Isn't' it ?

Friends , this is what the effect of surrounding on our mind ! Many people love to spend their lifetime in the company of good people , good culture and good surrounding !

But , can everyone able to receive such type of pleasant environment or surrounding everywhere they go for work or study ? Certainly not ! It is said that , after every 8-10 miles, the dialect of any language changes and people keeps speaking in different tone ! This speaking tone can be soft , harsh, rude, polite , arrogant ! Listeners has to understand the inherent meaning and motive of that communication and respond accordingly to yield co-operative output !

5.2 Typical Industrial Surrounding available before inspection :

Inspection Engineers has to accommodate and adjust with different types of thrilling working environment and surrounding ! While serving in such surrounding , they have to ensure , their decision of inspection stands correct and accurate !

There are some types of static surrounding where the movement of goods and services is not present at all ! You have to just visit the inspection site and record the daily observations ! Post installation regular inspections falls into this category ! Here you have to inspect the process flow and suggest correction if you observe any deviation to routine !

Such type of static set up can be any heavy assembly installed at work site , any installed pipeline , any installed recorder system , any system that allows or restrict material movement from one place to another ! Here these systems are functional from their point of installation and you just have to monitor necessary process sequence is being followed or not!

In dynamic form of surrounding, there is constant movement of products and services! This type of surrounding is observed in under manufacturing products work stations, various running machines, sites under construction, inside and outside in process work!

Here, you have to take care of available dynamic conditions. Things are well planned so that you can get your required time for inspection. There are set procedures and when the respective stage gets complete, you are asked to verify its accuracy! In online dynamic surrounding, there are multiple projects are operational from many work stations and you have to visit every station when the respective stage is ready!

So, depending upon the nature of surrounding which can be static or dynamic, your approach to inspection changes! In static surrounding, you get ample time to record your observations and derive inference of acceptance or rejection! In dynamic surrounding, you have to arrange your inspection tools smartly. In most of those cases, you need to be ready with special jigs and fixtures with the help of you can ascertain the accuracy of parts just by comparing with that special jig or fixture ! The characteristic feature of any jig or fixture is it is equally applicable to one part as well as million parts produced by that geometrical feature ! So, as far as inspection accuracy and speed is concerned for dynamic surrounding inspectors need special jigs, fixtures, sensors and measuring devices which are fast enough to record the necessary observation !

5.3 Physical Attributes of Inspectors :

What happens to a person when he is shifted from colder region to hot region ? What happens to a person who is shifted from dry region to humid region ? What happens to a person who is shifted from cities to villages ? What happens to a person who is shifted from open society to a restricted society ?

Friends, flexibility to available inspection surrounding is the most important attribute of every inspector engaged in static or dynamic surrounding ! Human body , if specifically trained will take its time to adjust to surrounding and once it adjusts itself, the regular job of inspection can be easily done !

Second attribute required involves technical knowledge of all types of surrounding ! When you are serving in corrosive environment , you must know the basic protection required before taking a step ahead ! If you are serving fast paced line production surrounding , you must be aware of causes of process link breakage and possible rundown or shutdown cost ! When you are serving remote work sites located in hilly terrain , you must be aware about the typical lack of resources in jungle prone area and hence daily planning required to do work with availability of all required resources !

Comfort is the thing of city ! In cities , you can get many resources easily . But when you are serving remote sites , gathering resources become extremely challenging , as an inspector ,you must able to manage your resources accordingly !

5.4 Mental Frame of Inspectors engaged in various surrounding :

Have you noticed typical schedule of a project manager taking care of 50 work sites for its quality requirement ? They may recruit 50 inspection engineers at 50 different work sites ,but they have to manage all 50 sites by sole decision-making skills ! The skill of inspection required at this phase is descent communication with other 50 inspectors and guiding them about accurate recoding and reporting of observations ! They also have to see the possible reasons of any safety risks !

When they are managing 50 project sites , they have to visit these sites and surrounding at all sites cannot be same ! One has to plan their two international visit, 5 domestic visits and rest of the inhouse inspection management visits ! So , the time availability of proper planning is very very important and inspector must able to plan their time table of inspection at different spots and sites accurately ! Secondly , within available time , they have to carry out inspection and give the decision of acceptance or rejection ! Hence accurate guess of stage readiness is also crucial !

What will happen if inspector notice major errors in at least 3-4 locations out of 50 work sites ? They have to keep track of the errors and note down properly ! When the rework will be completed , they have to visit the work site for re-verification !

5.5 Risks Occur during inspecting in different thrilling surrounding :

The biggest risk visible in inspection of different surrounding is availability of timely visit to that location ! When you are travelling to different surrounding , the accurate communication between stage readiness and your availability is important so that the production processes will run smoothly ! Inspection visits are arranged on the basis of prior notification in the form of inspection call ! When call is raised , it means ,internal inspection engineer has seen the stage and after conformance verification , he is raising inspection call for same ! Here as a respondent , you have to look your schedule and depute your available inspector ! If you are not able to depute inspector on right time , then the stage will remain idle and nothing can be done there ! Such idle time is not good for plant productivity and hence arrangement of inspector has to be done on time !

5.6 Challenges of different surrounding :

When you are inspecting in different surrounding , you need to be aware about surrounding weather , temperature range , network of similar type of manufacturers for suitable manufacturing options , service level agreements and financial impulse of accurate and wrong decisions !

When you are inspecting in same surrounding for number of years , let's say 10 or 20 odd years , the adjustment to that surrounding is not a big task ! After initial struggle of two -three years , you get habitual with your work way and then rest is increasing your responsibilities year by year !

When you are constantly exposed to different surrounding , you are not stable and you are constantly engaged in learning something new ! This imposes the basic discipline of upgrading yourself regularly as per the need of that surrounding and hence there are challenges of continuous training and development ! Some surrounding will boost your performance level while some surrounding will challenge your potential level ! Some surrounding will allow you to take your time but some surrounding will press yourself to provide the decision in minimum possible time ! In some surrounding , the rate of rework may be more and you need to arrange training to the operating people to reduce rework ! In some surrounding , there will be no error and your all efforts are identified by your pro-active steps before start of every new model ! Only when you systematically plan for new product quality , then only you get accurate jobs and hence reduction in rework ! Inspectors starts their career as early as 23 years and retire from the field at the age of 60 years ! In this span of nearly 37 years , they witness transition of three to four generations ! Earlier they enter as trainee and learn from other seniors and when they are about to retire , new trainee engineers are available with them ! This cycle of nature is the best surrounding !

THRILL 6 : THRILLING INSPECTORS

Image Courtesy: Sherlok Holmes , Pixabay.com

6.1 Introduction :

Till now , we have seen , various factors that affect the type of inspection ! In this discussion , we will spread light on peculiar findings by various inspectors under the title – Thrilling Inspectors !

6.2-Dimensional Specialist :

Do you know, dimensions play vital role in any product or part construction . If the dimensions of the part are not correct ,then final assembly of product will not match with interconnected parts and hence big dimensional mismatch error will occur which will need major rectification !

Dimensional specialists are well aware about very very critical dimensions and they check those critical dimensions as early as possible ! What does mean by as early as possible ? How can a part dimension be confirmed if the part is yet not manufactured ? Is this possible ?

Yes , critical dimensional verification even before start of part manufacturing can be done and the process is known as drawing review or design review ! In design review, apart from part dimensions, its safety and performance parameters and limits are checked against required minimum performance threshold while in drawing review the accuracy of drafting is checked with respect to design calculation !

Design calculation will calculate volume of 100 mm x 200 mm x 50 mm box by noting the weight of the part and its density while drawing review will show the part dimension as 100 x 200, 200 x 50 & 100 x 50 in front view, top view, side view ! In this way you co-relate the design calculations and designed dimension on drawing ! This is standard practice !

The dimensional specialist always refers to design data for which the part is designed. They note the minimum design thickness, design metal temperature, maximum allowable stress, heat treatment requirement and accordingly they calculate dimensions for that part by applying necessary formulae ! Dimensional specialists are not less than designers, they only perform the job of verifying the accuracy of that job ! For this work, they need to be aware about each and every dimension indeed ! Isn't it ?

As soon as they find an error in design calculations or pre-release drawing , they highlight that error and avoid it from happening with the help of design modification request ! Design engineer review this request and if it is correct , they amend the design and drawing and thus one error get corrected for that complete series before a single part is produced ! This special skill of dimensional specialist is known as drawing and design review skill ! So , if you want to see a thrilling inspector , just find out , how many design modification request he has filled for noting errors in drawing or design even before product production starts !

If these errors are noted later , then the cost of that rework could be very very high and it will be applicable to whole series manufactured under that particular drawing number ! Hence dimensional specialist looks for available part symmetry , their geometrical stability and proportionality , presence of suitable play or margin for future modifications , extra material provided for adjustment of critical dimension, profile verification with respect to male-female type of interlinking mating parts ! With this study the feel of dimensions becomes easy ! Along with this verification of drafted dimensions and dimensions shown in bill of material is also need to review to purchase right size of required material ! A small error in thickness or size creates purchase of wrong material !

6.3 Test Specialist :

Can you tell, a typical tube plate will show major water leakages even before a tube is inserted and welded to tube plate ?

Can you tell, the poured metal will have lots of porosity and blowholes even before the metal is transferred for pouring ?

Can you tell, the bore gauge will not pass through this tube even before the gauge is placed near the tube under inspection !

Can you tell, as soon as part is quenched in quenching bay, surface cracks will form immediately ?

Friends , this is the experience and knowledge of test specialist ! Even before a process starts , just by looking at its stage preparation , they could guess , what type of defect will be created in the final product and so what can be pro-active remedy so that such type of defects can be avoided ! This experience of skilled inspector saves crores of rupees of any organization in a typical financial year !

This experience comes after observations of more than 1000 tests of same type ! Let's say , if you have witnessed 1000 hydraulic tests in consecutive 5years of service , you get fair idea about the root cause of leakages , the required edge preparation before welding , the extent of cleanliness required before welding , the pressure lowering and raising in typical cyclical way to see the frequency of tube leakages !

Suppose , you have reviewed nearly 5000 radiography inspection reports in your 5 years of service , you will be easily able to identify the depth of particular defect just by looking at its radiographic image ! This is because , you are familiar about the actual size of defect and its radiographic exposure ! You know the basic technique of viewing the radiography and actual status of job before taking radiograph ! Because of this practical shop knowledge , your interpretations of defects become accurate and your decisions becomes correct !

6.4 Code Experts :

Design and manufacturing codes and their bonding with inspectors is age old phenomenon ! The code experts are the most thrilling inspectors available in your surrounding from whom you can literally understand each and every reason of part design and field safety ! Yes , code experts mean part safety interpreters !

The code experts are consulted at various phases of part production ! In review stage , their guidance is taken to confirm the trending general arrangement of terminal points and overall capacity utilization of that project !

During intermediate shop floor work , their guidance is taken to establish most promising quality assurance plan and important stages of inspection . Inspector has different point of view and since they visit many sites and many shop floor , their experience in the regard of what is to see and what has to be critically seen is enormous ! And hence they help manufacturer to prepare most promising quality assurance plan which will take care of safety of work site !

Code experts are also consulted when the part is handed over to customer ! But here their role is facilitator of earlier work inspected ! Here they act as your conveyor of system and they assure to customer that the quality management system at said manufacturer is in compliance with the international requirements and they have personally verified it by means of pre-planned inspection calls ! Even though , if some error happens at customer end after installation of equipment , they co-ordinate with the manufacturer to streamline the assembly after carrying out necessary correction ! Code experts are process masters !

On international code discussion conferences , code experts share their critical process observations and make code further suitable and easy for interpretation for other manufacturers , inspectors , auditors and statutory bodies ! Code experts spread awareness about general safety of people and products and they promote safety working practices !

6.5 Rework Experts :

Many times , practical instances are observed in shop floor or work site , where the team is required to carry out major rework ! The extent of work is so huge and a certified rework procedure is must before moving ahead ! Who writes such procedures ?

Rework experts are special type of inspectors who are aware about typical manufacturing processes and handling major reworks ! They suggest path of minimum damage and for this solution , first they think how the major rework can be avoided by carrying out some type of external support to existing system !

If the external support is not resolving the performance issue , then they suggest to add additional connection or additional arrangement of new opening in existing system to cope up with performance issue ! If still problem exist, then they suggest, easy way of removing joined part so that main system remains minimum altered ! With this suggestion the 98% problems get resolved and if there is unresolvable problem by rework, then the team has to manufacture a replacement for wrong job ! But till the decision of new replacement, rework specialist put their 100 % contribution to save the mammoth job from any defect or major repairs !

6.6 Influencers and Inspiring Inspectors :

These are most lovable people on shop floor or work site and they are supremely efficient in getting work done just by using their influence of experience, job knowledge, practical approach and other human values like better listening, logical questioning, motivating and inspiring !

People naturally co-operate with such type of inspectors and very less errors happens in their shop because they act with so much innocence and friendliness that all people follow their advice ! They are thrilling influencers !⊛

THRILL 7 : THRILLING CUSTOMERS

Image Courtesy: Shopping , Pixabay.com

7.1 Introduction :

If you are working in a multinational organization , you will come to know about the typical business concept of ' Customer Satisfaction ! ' Isn't it ? Customer satisfaction is important point of observation in business dynamics of such huge corporations and this is because such type of organizations serves to different type of customer's same time ! Every customer is business contributor and he is priceless ! In this discussion , we are spreading light on major variety of customers observed in typical business to business environment !

7.2 Traditional Customer :

Working with your traditional customer is extremely thrilling because he is giving you regular work orders and because of which your company is growing rapidly and you are achieving your professional growth rapidly ! Isn't it thrilling ? Your traditional customer will believe in your capacities even though you don't have orders from other customers because of difficult business environment around !

7.3 Novice Customers :

These types of customers visiting your company very first time and hence there is great thrill about how your professional relation will get develop in long term basis ! In business , you can receive order no 1 by your marketing tactics but for order no 2 & onwards , you have to rely on your product specification and delivered quality !

So , with Novice customers , you become familiar by understanding their job requirement and presenting your systematic achievement ! You also assure them that their all design ,production , material and quality requirements will be fulfilled from your side and your facility has already delivered such type of units before ! While working on Novice jobs from new customer , you may need to install new set ups, new jigs and fixtures , new inspection plan and in all these developments , your team improves as an efficient operational team !

Second thing , since the job is Novice , many people don't like to do it first time as it consumes more time than standard one ! But one who accept new jobs , invest their production time in its development and when that model gets totally develop , they produce that job in huge proportion and earns extremely fast profit ! For this reason , Novice customers act as important business element which test your capabilities and when they get proved , they also provide you reward in the form of bulk new orders of that developed product which others can't get !

7.4 Low Profit Margin Customers :

Such type of customers are very very good for newly found organizations which are waiting for new work orders ! Initially very few people believe in your capability and in this phase, you need some starting orders before you actually put-up things in actionable mode !

In low profit margin the technical scope of work is very very limited and you either required to carry out one of two technical conversion operations before you handover the work to your customer!

Such activities can be making bulk production of small and easy component such as plugs , rivets , nuts ! Here per piece profit is lesser but scale of production is higher ! So , if you are able to produce more , it adds to your profit kitty and thus your business gets a start !

There can be services like cladding or painting where you have to apply two or three coat and you get good amount of money for that work ! If you satisfactorily do that job , you get more orders and thus earn a descent profit !

7.5 Defense Sector Orders :

What can be an inspiration when you get work orders from your national defense department to produce parts that will contribute to national security ! This is an extremely thrilling moment for every manufacturer and designer to present his well-designed part to defense officials , seek their clearance and then start product manufacturing ! Not only this , the moment of supply of first batch of product duly inaugurated by respected defense official is another icing on cake moment !

While handling such type of orders , you understand the importance of National security .

You observe the factor of safety expected in parts . You observe the material class and its quantification so that nothing will remain short ! You will see the instructions given before production and how the certification of completed jobs to be done ! There orders will also make you feel that although you are handling defense sector job but still defense sector put great confidence in your capacity and hence you have free hand to implement the most sophisticated innovation to make defense system stronger and stronger . Such attempts will be suitably appreciated and rewarded in right social forum !

7.6 Complex Job giving Customers :

In a dynamic business environment , there are people who want to create something special and different ! However, their expectation is not fulfilled by routine manufacturers ! Routine manufacturers like to do business in their own 'status co' position and hence such customer find out manufacturers like you who deals with complex job easily ! As a reward , you get premium quote for your creativity and thus your business niche excels to new horizon !

Basically, such type of jobs comes to you after rejecting by number of other business competitors ! Really ? Yes , really ! Your other business competitors look towards project drawing and look towards the available work in hand and simply reject the job by telling the reason of sufficient load availability at that time ! Actually , there are few process steps where manufacturer has to install special provisions and because of which they are not ready to carry on desired investment !

Then this job comes to you . Your shop is comparatively active and you host many provisions in your shop with the help of which any type of job order can be suitably accepted ! You see the drawing , you see the workload , you see the extra team engaged in supporting regular team and you accept the order !

A clever manufacturer always keeps one team extra for such type of special and complex jobs . When regular work is there , he breaks up that team in small parts and ask them to engage in regular jobs . When complex jobs come , they reorganize that extra team and provide them their pre-determined tasks ! With this shop floor management strategy , people could see they are handling a different job avoided by their other competitors and hence in challenging business environment , they can get upper edge when this order will be completed . Again, other customer will also approach your side and thus your business momentum will increase by accepting complex job orders !

7.7 Semi- Finished Jobs Customers :

This is a very very special class of customer and if you receive such type of order , you receive a higher amount of profit than same job of such type ! Why ?

Such orders are received to you because of some business exigencies ! Exigencies can be any , they need not to be mentioned but in the end the final decision taken in the business is shifting of partially completed job to different manufacturers by accepting the responsibility of completed work ! In the job folder the scope of individual manufacturers in fairly mentioned and all responsibility of product acceptance is owned by customer itself ! Customer just want that job to be completed to reduce his losses because of incomplete burden on their financial system !

When the order is given , it is said that , the following activities are completed and its technical records are acceptable to order provider ! The balance scope of activities is like this and this will be quality assurance plan for balance stages . The new order full filler has to complete the balance task and show the completed status to order giver . On satisfactory completion of this order , technocommerial bills will be settled ! If the manufacturer has any doubt regarding current status of the job , they can raise technical queries , which will be discussed and resolved before handing over of fulfillment order of this job !

With this scope clarity , manufacturer receives part drawing , balance material , incomplete job and his team starts completing the remaining work ! The product's final certification is order providers responsibility and the name of manufacturer is mentioned as one who starts that job ! This way the second manufacturer bear the responsibility of fulfilling incomplete order as per their customers quality assurance plan ! The name of second manufacturer can be added as joint manufacturer if records specifically demand it ! In any industry , any work is processed at two three different place and acceptance at all such places is carried out with respect to quality assurance plan !

7.8 Urgent Customer :

Currently every customer has urgency but that urgency is made for early completion of their orders . However, there are real urgent customers who want to finish their work in least available time in their hand and for this work they are ready to pay you price which is minimum double or triple time your regular quotation ! If you give this job , they may save huge penalty and other payouts and hence they need a fast and accurate executive ! A prudent manufacturer is always ready to accept such urgent order as he know this is real service and chance to test his systems peak performance!

THRILL 8 : THRILLING AUDITORS

Image Courtesy: Auditing , Pixabay.com

8.1 Introduction :

Audits are part of regular system verification to see whether the documented system manual is enough to ensure the quality assurance and safety of that system and surrounding ! The basic purpose of audits always revolves around measuring effectiveness and efficiency of the system !

There are different types of audits and different personalities are involved in audits . When organization is smaller, one or two auditors can review the system and declare audit findings and observations . When the organization is huge , a panel of auditor visit the organization under the leadership of the lead auditor ! Individual auditors perform audit of assigned section and the collective reporting is shared with lead auditor who evaluate the audit findings and accordingly issue certificate of audit compliance or suggest action on observed non-conformances !

In this discussion , let us see , how many auditors visit your organization and how you report them during that interaction ! This knowledge will help you understand your system in much simpler and better way !

8.2 Statutory Auditors :

Number of statutory auditors visit your premises after formal application and filling up prescribed fee for that audit ! Statutory audits are carried out for granting you typical government permission to carry on the intended business activities . Statutory audits are also carried out for renewal and recertification of your system after pre-determined renewal schedule !

To pass in this type of audit , all necessary document requirement is available on dedicated government website and in case you fully comply to that system norms , officials grant you certificate of approval !

The basic purpose of every statutory audit is to ensure public safety and review whether any illegal activity is not performed in the premises which will alter the purpose of issue of licenses for particular business purpose ! If such major deviations are observed , then statutory auditors' issue authorized action report where legal penalty and terms of dis approval of particular license are involved !

If there is partiality in any statutory audit , the audit applicant can request for justice in suitable judicial forum or court through their legal advisors !

8.3 Quality Auditors :

Every organization is required to run as per typical quality assurance and control system . Every production process has typical international standard which provide detailed guidelines about part manufacturing , purchase , testing , design , sales and marketing , servicing !

The role of quality auditors is reviewing your system manual which incorporate several system standard and responsibility matrices . To every responsibility holder , they ask typical audit questions after observing one or two system generated products ! The manufacturing method and documentation of manufactured product must match with each other ! When this identification and traceability is properly established , it reflects system and surrounding are in compliance with each other ! If any deviations are noted , then relevant nonconformance report is issued and detailed root cause analysis , corrective action and preventive action is suggested to again comply with the system !

In a typical quality audit , auditors visit a particular working area and observed which process is going on there . Then they check whether process chart , observation recording register is available . Then they check the qualification of person employed on job and they ask questions about organizations vision and mission statement to that person to check the flow of top-down communication !

After this verification , the documentation of every completed and in process work is checked and the compliance to observed versus specified written performance expectation is verified ! From these observations , the overall monthly completion of stages and their accuracy is verified and accordingly systems efficiency is calculated . From this calculation , how much effectiveness in the system is derived and ways and means of effectiveness improvement may be suggested in ' Areas of Improvement ' segment !

Because of quality audits , your all quality documents, reports and records are audited and remarks regarding compliance or non-compliance are given . Errors do happen in working environment but there has to be written procedure to deal with error and ensure part safety after error correction and its documentation . Because of this full proof approach , quality audits are considered as customers faith builder ! When a system is recertified by quality audit after period of three year , it reflects the work done in that system is systematic and people are quality conscious !

8.4 Safety Audits – Health , Safety , Environment !

Safety is prime most requirement before anything else ! When the working environment is safe and secure , it is affirmed that system is running as per laid done processes and there are nil chances of any unpleasant safety related incidences !

However, in practical conditions, during various types of work requirement, there are chances of happening of some safety related incidences, some near misses and some accidents which are not desired in any professional organization ! Safety pointwise, every accident can be avoided and human efforts shall be taken in a way that assures total field safety to working and visiting people !

The process of safety audit focuses on safety aspects. Which consist of identification of various work area according to their natural freedom for movement ! Some area has public access, some area has special access while some areas are prohibited and only certified people are allowed to enter in such area !

Tagging and identification of hazardous and non-hazardous class of material is another important aspect of safety audit. Material has to be safely stored in dedicated area and it has to be safely discharged from system after its intended use !

The written procedures of safe handling of equipment's, material and machines should be available in visible area and warning signs has to be indicated in potentially risky area !

The severity class of every possible risk is identified after noting its classification and accordingly risk mitigation plan has to be followed !

Daily incidence sharing and tool box manual apart from system maintenance manual need to be ready with relevant details so that one can easily glance through past or present status of safety !

Data of safety can be analyzed with respect to system efficiency ! It's natural that when there are no accidents , system efficiency is higher ! When there are instances of accidents , necessary corrective and preventive actions are needed to be taken . If these are taken quickly , rate of accidents reduces , if those actions are not taken , similar accident is faced by different operator !

Shop safety index, site safety index , material handling safety index is verified before certifying compliance of safety system ! Safety auditors also note the use of PPE and modern surveillance system to take care of your system on day-to-day basis !

8.5 Welding Audit or Process Audit :

These types of auditors are specialist of particular process and they are well aware about particular process deviations done in the shop floors or work sites !

Process audit like welding audit is carried out in establishments where welding is a predominant process of manufacturing and other processes are supplementary !

In welding audit welder qualification , WPS review , PQR review , weld consumable and their test certificate review, weldment destructive and non-destructive testing records , weld defect records , welding improvement projects , weld maps , weld edge preparation methods and all available types of welding processes are seen and their compliance is verified !

When welding is pre dominant process , frequent welder qualification and requalification is necessary to be done ! Auditors do find some deviations in the welder qualification record and ask manufacturer to requalify the welder before allotting to job work ! There are serious penalties to manufacturers in case he found engaged in activities that affect final welding strength and quality adversely !

8.6 Financial Auditors :

The balance sheet , account , ledgers , bills , request forms , financial approvals and sanctions , budgets and provisions , benefits and perks , employee salaries and material purchase cost , overheads and fix cost , all such financial indicators in a system are verified in financial audit !

Financial audits reflect the transparency of any system . How the funds are used and how the record of every single rupee is maintained on long term basis! The overall tax credibility of any organization is observed with financial audit ! The position of assets and liabilities and hence overall financial stability and liquidity of system is verified its legal financial compliance !

THRILL 9 : THRILLING DEFECTS

Image Courtesy: Industrial Insurance, , Pixabay.com

9.1 Introduction :

The basic purpose of Engineering Inspection is to assess the conformity of products or services to laid down specifications, standards and codes ! When you are assessing conformance , your main motive behind this action is to look out for any Non-conformance ! Non -conformance are classified into two types ,one is discontinuity into regular structure and second one is defects observed in regular structure ! Every defect is a discontinuity but every discontinuity is not a defect ! Discontinuity can be repaired by carrying out approved rework procedure , however defects are meant for straightforward rejection !

In this discussion , let us understand the thrilling defects observed during typical engineering inspections .

9.2 Alignment Defects :

This type of defect is noted when the mating parts are not fitted correctly and their connection doesn't fall in required geometrical plane which can be horizontal , vertical or inclined plane ! Mis-alignment create faulty connection liable to various types of gaps and leakages !

9.3 Missing Fitments :

This is one of the major defects noted in big assemblies ! Big assemblies consist of more than thousand small functional parts . These parts are separated into smaller assemblies and you need to ensure that every assembly is complete in all respect . If there is any missing fitment , the intended operation of the system gets stuck and when you review your assembly from start to end , you notice the missing parts ! When you fit them at right place in the assembly circuit, the circuit start functioning and you then certify that assembly as a complete assembly !

9.4 Incorrect Orientation :

Incorrect part orientation is another common defect noted in mega assemblies ! Orientation refers to left hand side view , right hand side view , top view and front view !

When you fit a part in the assembly , you have to check its orientation by referring to front – top and side views , in some cases you may need to review its bottom view and rear view ! In all such views , you need to ensure the desired orientation is accurate as per given direction and it is not fitted reversely ! If you fit it in opposite direction then after assembly the part will not join each other which will create major error in further operations !

9.5 Connection Errors :

In any assembly , there are number of connections present to carry out required flow in the system ! Every connection has a typical design size and its further relevance to assembly line ! Suppose in a typical piping connection , you can have pipe ranging from small size to large size . These pipes are connected to each other by flanged connection . Open ended pipe is connected to flange -pipe connection known as Nozzle which is welded to main job ! If pipeline to be joined by nut-bolt-gasket-flange connection is observed with small sized flange than required , then this connection could not happen ! Simlilary , in any electric circuit , if required part rating is not maintained at required location , then the sufficient amount of electric signal could not flow ! One has to ensure , the flow of system transfer through well placed connections !

9.6 Plate Defects :

Plates are major construction material and they are used to build major covering of the equipment ! Plates are manufactured in still mills and during manufacturing , some type of plate defects are observed which can be classified as plate lamination , pinhole and porosity , edge cracks , blisters , scaling , pipe , composition deviation ! These defects are not good for structural integrity of main product and they need to be removed before plates are dispatched to designated customer !

9.7 Pipe & Tube Defects :

Another engineering shape which is used for fluid transfer is Pipe & Tube !

Pipes can be seamless or electron resistance welded – ERW ! Tubes can be wrought or as cast ! The typical defects observed are related to process of their manufacturing ! When tube is manufactured by centrifugal casting process , the major defects noted are thickness variation , cracks , diameter variation , insufficient material fill up , porosity , blowholes , shrinkage !

ERW tubes and pipe can experience opening of weld and subsequent leakages . There can be waviness and out of roundness in some tubes and pipes ! Pipe and tube identification nomenclature to be ensured correctly ,sometime ,deviation is noted in material identification !

9.8 Process Defects:

Engineering inspections is vast field and it consist of involvement of number of processes to create a desired product ! So, according to processes you can have rolling defects , you can have machining defects , you can have casting defects , you can have welding defects , you can have forming defects , you can have plastic molding related defects , you can have bending and shearing defects ! To avoid such type of defects , you have to ensure , your process parameters are accurate and as per your prescribed machine setting plan ! When your processes are running swiftly , less defects occur in this system ! When there are issues in process , until that issue get resolve , the defects keep occurring !

9.9 Painting Defects :

Paint and adhesives are used to protect the surface of engineering equipment . If the protective coating has typical required dry film thickness , then the surface underneath remains protected ! Painting defect involves less DFT , wet paint , paint spill off , paint not joined with primer , shade difference , heavy coat and light coat visual deviation , paint blisters and powder clogging , brush marks and excessive thinning of paint !

When painted parts are installed in work sites , the natural wind and climate present there start corroding the assembly . If the paint is stronger , then it resists that corrosion for long time and if painting is not according to specification, then first paint gets removed and then the surface start weakening because of climate induced corrosion !

9.10 Cladding Insulation Defects :

Aluminum and stainless-steel cladding is used to protect the layer of thermal insulations . Thermal insulations safeguard your external system from your main heating equipment from its excessive skin temperature ! Because of thermal insulation and supporting protective metallic covering which is known jointly as lagging and lashing , the surrounding remains cooler and heat is insulated from reaching external environment by thermal insulation !

So , such type of insulation can be loose fitted , some slabs may not be fitted and a gap is remained there which can heat that part excessively and may create risk of overheating , the cut out can be fitted loosely and riveting may not be done , part name plates may be fitted in opposite direction which alters the view of part observation ! For this reason , you have to carefully see the compactness of insulation by striking typical hand flow on it ! If the insulation thickness is correct , your hand will not go deep down the main surface of equipment , if the insulation is not filled , your hand blow will reach the direct surface beneath the cladding ! This is serious issue and if proper thermal insulation is not provided for part , then there is risk of injuries and burns to operators working in that area !

9.11 Assembly Defects :

In assembly , you have to fit part in particular sequence . The output of one system is input of other system and thus intermediate gaps are to be filled with designed controlling mechanism . If input- process control -output is fitted correctly , your system serves intended function . In case there is an error in assembly , the final outcome become wrong !

Mechanical assemblies are created in shop floors and they are fitted in site foundation . Here site foundation and your mechanical assembly must match with each other. If it doesn't match, then you have to adjust your foundation accordingly or if possible, you have to correct your mechanical assembly base with respect to corrective drawing ! The line- length- height mismatch has to be avoided in assembly !

9.12 In service Defects :

These defects are noted after installation of equipment's and they are attended with respective site visit ! There is typical instruction manual provided for safe operations . These defects occurs when non-standard material is used as input and which affect internal mechanism over the period of time !

9.13 Defects noted in destructive and non-destructive testing :

These defects are related to structural integrity and material quality . Low tensile strength, lower elongation, cracks in the bend test, porosity, blowholes, slag, are some of the typical defects ! ✳✳✳

THRILL 10 : THRILLING REWORKS

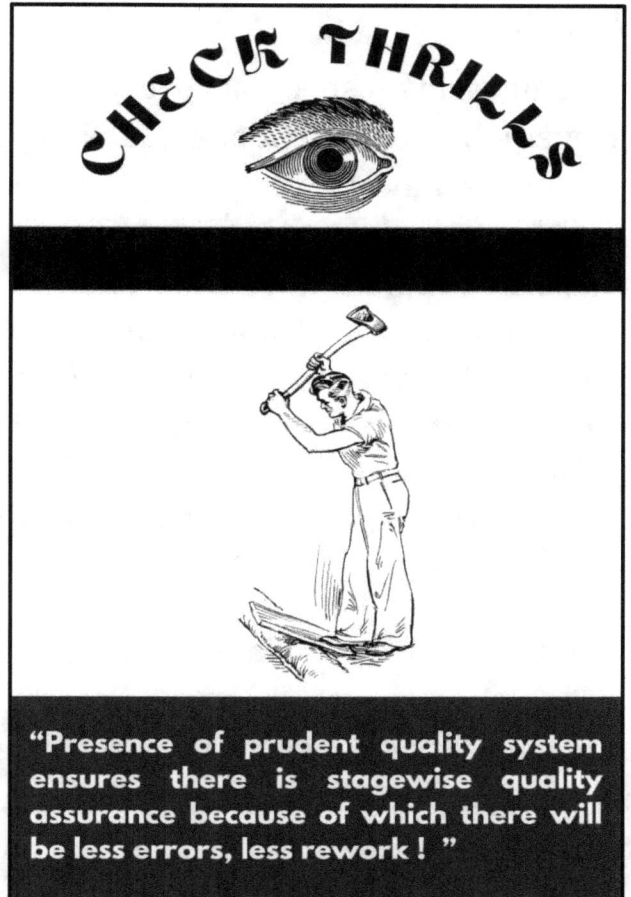

Image Courtesy: Axe , Pixabay.com

10.1 Introduction :

In last chapter , the thrilling defects are discussed , in this chapter thrilling reworks will be discussed to get the idea behind thrilling inspection !

Engineering Inspection is very very tricky and clever job ! Competent and qualified people carry out technical work with respect to approved drawings and age-old experienced procedures . People think , there must be least chances of errors in work done ! For some extent its true but in big extent, this is not the case ! Else why the second or third big department in any Industry is known to be that of Quality Assurance & Control ? This is the practical need of having QA department in your organization !

If you wish to remain for long term in business environment , then you need to have quality and safe products for your customers at economical prices ! Presence of prudent quality system ensures there is stagewise quality assurance because of which there will be less errors, less rework and fast paced accurate operations reducing the overall down time or idle time of your machine and hence you can improve your productivity and system effectiveness !

Let's see , how many types of reworks are performed in industrial manufacturing environment !

10.2 Allignement Rework :

The exact amount of mis-alignment is measured with scale and similar profile measurement instruments ! Accordingly, the surface is reshaped to desired symmetric profile so that both joining faces touch each other without creating any gap or leakage of any type ! When the both surfaces are aligned correctly and a fastener like gasket or spiral gasket is placed in between and the arrangement is tightened with nut-bolt -washer hardware , the joint become water tight ! You may take hydraulic test by filling up water inside that vessel and see whether water is coming from aligned joints !

There is spirit level check carried out where the spirit bubble remains at the center when the surface is properly aligned !

Surface preparation for right alignment is carried out by suitable grinding, machining or polishing process depending upon the nature of job!

10.3 Missing Fitments Rework :

In this rework, you have to first remove the incorrectly fitted part if any and then you have to carry out necessary edge preparation so that the new part can be easily joined. You have to check the change in available distance because of rework and same has to adjust in new part fitment. If this adjustment is not possible, then you have to modify the approved drawing and ensure the right joint length is included and new material is used for that particular joint combination! The major challenge of such type of rework is reduction in joint length because of incorrect part removal and hence you have to take new part with suitable length! While carrying out this rework, the edge preparation needs to be accurate to avoid leakages of any type in reworked assembly. In case, leakages are observed, then suitable stringent process to be used which assured leak proof joint!

10.4 Incorrect Orientation Rework :

When you highlight this type of defect and ask your team for its modification , if the defect is observed in set up stage , then the team has to remove the temporary tacks and re-orient the connection in right direction . In case the defect is noted after permanent welding , then they have to cut that joint by suitable cutting process such as gas cutting or hacksaw cutting , lathe cutting , grinding and again make edges suitable for joining !

The orientation can be changed in between 0-360 degree or in some different plane , you have to note the accurate fitment and join it correctly ! Once orientation is corrected , your subsequent assembly also need to match accordingly !

10.5 Connection Errors Rework :

This is tricky rework ! First you have to see , what type of connection error is happened ! If the opened hole is smaller than required hole and it is noted in the inspection , then you have to increase the size of the hole and fit the right pipe or right part for which that hole is made !

If the opened hole is bigger than required hole , then it create worries ! No issue , you have to see the possibility of shifting that connection to nearby available place and make changes in mating piping connection ! The opened hole is blanked by making a dummy connection which is blanked by mating flanges and it is not used anytime in actual servicing ! The new connection is opened in available place on same co-ordinates and the joining is done referring to approved drawing dimensional adjustment scope ! Suppose a connection is shifted by 200 mm and its size is increased by 100 mm , then you have to fit a mating blank nozzle at this spot and you have to open new connection of required size as per drawing dimension !

Same adjustment to be done for circumferential shift . You have to blank incorrect connection by using blank flange and you have to open new connection at required X-Y -Z co-ordinates !

Same kind of rework is done for other parts where axial dimensional displacement happens ! You have to cut that incorrect joint and align the job to required symmetry and then join that portion correctly !

10.6 Raw Material Rework :

When you find defects in purchased raw material during its receipt inspection or during intermediate processing , they immediately contact your supplier and show them those defects before clearing their purchase invoices ! This is very very crucial stage !

In this case , if your defect observation is correct , supplier immediately plan for correct supply and accept defective lot at his end at his cost ! In some contract , free of cost replacement is assured in case wrong quality supply is provided . In case , material defects are noted during intermediate processing and if by removing that defective part , rest of the part is satisfactory , then with consultation to suitable inspection agency and supplier , the defective part is cut and its cost is reduced from original bill ! In case , defect is noted in extreme last phase of testing , then free of cost replacement along with agreed delay penalty has to be given by supplier !

10.7 Process Defect Rework :

At the end of every process , suitable edge preparation and profile preparation is carried out ! Rolling defect are removed by re-rolling parts ! Welding defects are removed by removing that defect by in-depth back gauging process and then refilling the weld joint by multilayer welding ! Surface cutting defects are removed by grinding sharp notches and making surface smooth ! In some critical joint , material edges are prepared by machining ! Machining defects are removed by following surfacing methods . In NC & CNC machining , accurate profiles are achieved by accurate programming ! Defects observed in heat treatments are very different . If there is permanent deformation , then that defect cannot be removed ! Hence heat treatment cycles need to be critically monitored with respect to progress graph plotted against time -temperature and applicable thickness ! Defects noted in leak testing are removed by draining the water , taking part on workplace and then reaching till the depth of defect and then filling that defect with suitable joining process !

10.8 Painting Rework :

Painting defect can be visible in stationary area or on entire surface ! If the painting defects are present in local region , the local repainting is done by keeping remaining area under proper cover ! If full part needs to be repainted , then it is shifted to painting booth and again painted till required dry film thickness is achieved !

The rework procedure of painting involved surface preparation of required grade of surface roughness which can hold paint particles tightly ! Then applying primer and checking its thickness ! When the first coat is dried, then final coat is applied ! Dry film thickness , painted surface texture is observed critically to provide required matt or glossy finish !

10.9 Cladding – Insulation Rework :

This type of rework consists of preparing correct cut out and fill up the required thermal insulation ! This type of rework requires proper tightening of rivets and screw with main job and there should be no gap so that water or other fluid pass through insulation !

10.10 Assembly Rework :

The falling & overlapping part of assembly are suitably positioned in this rework . All arrangement is done as per sites general arrangement and terminal points drawing ! Rating and identification of all mechanical parts are noted before dispatching that assembly as it is or by dismantling !

10.11 In Service & Lab related Rework :

Noting the site environment and adopted practice , the observed issues are resolved with site action report ! About the destructive and non-destructive type of rework , the retest and repair -retest formulae are applied before certifying part under inspection ! ✹✹✹

THRILL 11: THRILLING APPROVALS

Image Courtesy: Check Box , Pixabay.com

11.1 Introduction :

Approvals are permissions given after satisfactory verification of technical and managerial as well as statutory requirements of any and any product or service under consideration ! When things are approved ,it means they are safe for the purpose they are made for and made with !

In this discussion , we are going to spread light on many engineering approvals which permit certain type of work ! In case of disapprovals , the approving authority issues reasons of not approving the part or service and the applicant has to rectify that objection with detailed technical analysis and if there is any mis-conception by authority itself ,then the way of fair technical arguments with respect to technical standards ,specification and code is always available till the matter is correctly approved !

Every approving authority is partially or directly responsible for loss of damage that happens because of approving faulty parts and inferior services . They have total freedom and authority to reject part and services which endanger the bio-diversity and basic public health ! Still , if they are approving the part or service may be because of ignorance , over confidence , illicit means or any reason thereof with which the justification of approval is technically not satisfactory , such approvals are liable to higher level legal scrutiny and subsequent legal punishment in case of any unpleasant minor or major safety related incidence ! Hence , one has to take utmost care before approving any document , any part or any service !

11.2 Drawing & Design Approval :

This is the most crucial approval of any engineering work ! Engineering drawings are schematic and detailed representation of overall assembly of product in the premises where product or service will be served !

Here detailed layout is designed considering the minimum and maximum requirement of customer and its overall financial estimate ! In case customer is ready to pay any higher limit, then things are designed for long lasting impact . If the customer has tight budget , then designers try to offer best in class material in that budget or they may suggest alternative way with which same goal can be fairly achieved . This is customers call and designers are just supporting their field requirements in best possible technical way !

Here approving authority verify adherence to statutory requirements, technical specification applicable for that product or service , provisions made for easy facilitation , safety and emergency handling and overall security concerns of surrounding where the product will serve for long time ! When all these things are observed as per laid down guidelines by respective controlling body , the approvals are given to go ahead for subsequent execution of plan ! Approvals can be given for first job or for whole batch depending upon the innovation of that job . When job or concept is very very new , first part is observed as research and development unit and when all safeties are proved then next batch is approved for applicable serial numbers ! Type approvals and batch approval determine the overall productivity of any engineering sector !

11.3 Material Approval :

This is strength approval ! Engineering materials are used for excellent mechanical strength , hardness , wear resistance , fair amount of elasticity , impact strength , fatigue strength , creep strength , compression strength , malleability , ductility !

Material naturally possesses these abilities or with some metallurgical and material processes , some of the properties are modified ! Material standards are used to refer right material for right application . Depending upon the work environment , suitable material is suggested in material standards . Such standards are applicable in international perspectives and hence anyone can refer it !

In material standards , physical and chemical properties of material along with method of manufacturing and testing is given . Method of material mill certification is also given . If a typical mill producing material as per reference standard , then they have to obey all instructions and guidelines without fail to certify that material conforms to that respective standard . If any deviation is noted in produced material with reference to standard written in certificate , that material is liable to rejection after destructive and non-destructive testing results . Because ,whatever written in document is not observed in actual testing ! Such certificates are considered as misleading and hence such materials are technically not fit for particular intended purpose !

So , material inspector , refers standards mentioned in material test certificate and check every aspect of that material as per reference standard guidelines and mandatory requirements ! There are few supplementary requirements also , which changes as per case to case !

Material carries almost 60-70 % price of overall product and hence its approval is critically important ! When whole material gets approved , rest of the job remains is accurate workmanship to create that product using workmen's skill , knowledge and experience !

11.4 Workmanship Stage Approvals :

Depending on severity level of product or service, the quality of workmanship is approved with the help of suitable stage inspection plan ! Product manufacturing is a process and once material is approved for manufacturing, it undergoes different forming and joining processes to result into final finish product !

A product can have just one stage in between raw material to final product conversion or it can have as many as 100 stages in between conversion from raw material to finish good ! Depending upon product design, quality assurance people design technical stage approval plan in discussion with prime designers and prime shop managers with mutual job knowledge and need of mandatory inspection to avoid any undesirable defect !

There are few types of stage inspection and their controlling mechanism ! Stages can be verified, reviewed with reference to stage completion document or it can be a 'witness stage' which has to be witnessed mandatorily by inspector ! If he doesn't clear that stage because of any technical reason, then the stage is kept on hold till the rectification of technical concern highlighted by field inspector !

The observations of stage inspections are written in quality plans and check list and final decision of approval or rejection is mentioned . With this written communication , things move ahead ! When a job is certified by quality , it means , it meets all requirements designed for that job !

11.5 Laboratory Approvals :

When abnormal findings are observed in jobs because of any process deviation or process variation , the work samples are suitably collected and sent to specialized laboratories for their defect evaluation and root cause analysis .

When the observed root cause is verified in faulty process with the help of its laboratory findings, the responsible people have to amend process and again test the produced samples . If new samples are observed as correct , then that process modification is updated in documented system and followed subsequently ! If the process result is not changed , then multiple testings are done to find out the exact influencing parameters ! In case of multiple complex process defect , fields specialist work together to find out necessary corrective and preventive plan by in depth ,minute elemental level analysis !

Lab approvals can be applicable to one sample , one lot or whole batch depending upon the type of testing required !

There can be physical lab, chemical lab, material testing lab ! Laboratory means quality assurance certification by scientific evaluation !

11.6 Manpower Approval :

If material consist of nearly 60-70 % of product price , then the next investment done in deputation of required manpower ! Production manager know the ideal manpower requirement and accordingly depending on the current work load , the manpower is hired with verification of qualification , experience and approach toward work ! While hiring manpower , overall market scenario and orders in hand are studied. Along with this expected market boom or slack is also studied to depute optimum manpower so that there will not be any job cuts on large scale ! Manpower approval affect productivity of plant directly !

11.7 Facility Approval :

You are working in an industrial environment and industries basically work for delivering people friendly products and thus they generate profit ! The part of profit is approved for creating good facilities for both customers and employee from organizations side ! These approvals are taken at top management level and in many cases number of employees also share their work-related facilitation so that there is increase in productivity !

11.8 Financial Approval :

Financial approval is backbone of every approval discussed before ! When technical , managerial and statutory aspects are meet , financial approvals are given with funds ! ⊛

THRILL 12: THRILLING SITES

Image Courtesy: Danger , Pixabay.com

12.1 Introduction :

Most of the capital goods are dealt with business-to-business order booking method . In this method , authorized representatives of one organization award a purchase order to other organization who regularly manufacture that product and has good to excellent market recognition ! Further to this reputation , the manufacturing organization is certified with international quality system along with relevant upgrades of health -safety -environment certifications and hence purchasing from such well-known manufacturer is a reliable business affair of high value transaction !

Once the product is manufactured , its assembly has to be done on worksite suggested in contract terms ! A customer can have his central corporate office in city like Mumbai and he can have his worksites in any part of the country like Bengaluru, Hyderabad, Delhi , Chennai ! The manufacturers have to supply the assembly at address designated in the purchase order ! The technical look out is co-ordinated by team working at job site while commercial look out is handled by corporate office once assembly is handed over from manufacturer to customer !

So how many types of work sites are there and how they are served is the discussion theme of this topic ! So , let us look into number of work sites which are served by manufacturers for assembly of their part ! The need of inspection is mandatory at all work site before handing over project to customer !

12.2 Export and Domestic Job sites :

Work sites which are located outside your home country are known as Export work site ! Work sites which are located in your own home country are known as domestic worksite !

Apart from design , manufacturing and testing of manufactured part as per export customer specified and territory applicable design code , other care need to be taken in adhering to local governance rules and regulations ! Not a single hammering is allowed if that country doesn't permit such type of rework once sales is invoiced ! So , manufacturers have to take due care to ensure the supplied part is just a 'Plug and play ' type of a job and no rework will be done there which may require various permissions and approvals from international authorities !

In domestic work sites , you have to provide service in your own home country . So , your job can be sold in same state or it can be sold in some other state . As per the site's location , you have to depute your nearby servicing and commissioning team to assemble that product and handover to customer after successful in-service trial !

The coverage of domestic manufacturing facilities increases as manufacturers find more orders within his own country . For this type of business expansion , they open up manufacturing facility at major locations of the country from where they can supply products to desired customers economically ! Also, the response time for after sales service will get improve !

The main advantage of domestic sites are your site expenses are within your local currency limit and you don't have to spend foreign currency !

12.3 Urban – Semi Urban – Rural worksites :

It is well known fact that most of the industries are situated in major cities of your nation and hence maximum business transaction happens in the cities . One of the most important benefit available in the city is skilled manpower and resourcefulness coupled with excellent connectivity !

If an organization sells 10 assemblies in a city center and if they have to install that assembly at 10 different points within city , this task can be done by team of 20 people in just matter of three hours ! How ? The team will get split into combo of 2 teammate and every pair will access the assigned installation point and after installation they will hand over the assembly to their customers ! Suppose one assembly has value of 3 lakhs , the team of 20 people will do the sales of 3 x 10 = 30 Lakhs in just three hours ! This is the benefit of urban worksite . You can earn as much money as many orders are available in the city !

In semi-urban worksite , you have to first transport your job from urban location to semi urban location and then depute your team there for required installation ! If you have fairly developed your service network in all parts of your country, you can easily depute nearby team . If this network is weak, then you have to arrange people from urban place and they need to install those set ups in semi urban places ! You need to provide extra allowances for this on-site service and also there is fair amount of outstation travelling, lodging and boarding charges application ! You have to consider these expenses while providing quotation of your supply !

Rural worksites are comparatively very few but they do exist ! When an urban manufacturer opens up a manufacturing facility in nearby village to save municipal taxes and get the benefit of opening industry in rural area, suppliers have to come to that site and carry out the installation ! Secondly, industries where prime raw material is processed in rural area has their major operations located in rural area ! Just see the example of co-operative sugar mills, edible oil industries, various agro based industries, paper and pulp industries, fragrance & aromatic industry, metal & mining industries they are situated in rural area from where the raw material is easily accessed and processed saving huge transportation costs ! So , you have to access this area with your transportation team and carry out the installation of your products !

12.4 Regular – Risky – Critical Work sites :

This type of site classification is done based on the severity of site environment in which your product will be used for intended service . This type of classification is done to highlight the necessary site safety during installation and handing over !

Regular work site has quiet and easy-going working environment , here risk of life or any type of property damage is very very less and hence the site work is considerably limited ! Suppose you want to install 20 blowers in a big college for its cooling and air cleaning need , you can easily install ! Suppose you have to install 50 heaters in three star hotels rooms , its comparatively regular and less risky work !

In risky zone , you have to work in little bit hazardous environment and in that environment more than one high risk operations may be running on constant basis. So, you have to either wait till those operations get finish or you have to ask the customer for suitable separation from that environment till the installation is carried out ! Many industries where potentially hazardous chemicals and material is handled regularly comes into this category. Here ,before every installation , local safety officers' installation activity permission is must ! Under his observation , you have to carry out your scope of work and in case of any urgency , you have to take their support !

In critical work sites , even touching the territorial part is not allowed ! Why ? These are critical jobsites which are producing extremely important material which is useful for national needs ! These sites are strictly prohibited to normal citizens or any unauthorized citizen and access to these worksites without proper permission is legal crime !

So , before starting work at such critical worksite , you have to provide your all identity related documents and get the approval of local safety team so that you can work there for limited days period . In case you could not complete the work in allocated time , you have to request for extension of existing permission . If they allow , you can work , if they demand penalty for delay in work , you have to refer to your service contract terms !

When you start your work at such critical worksite , you have to maintain strict confidentiality of that worksite and you have to ensure that your work meets the laid down specifications and requirements !

In case of any causality during work , the local support team is all time available and you have to sign up respective forms in the beginning itself so that you can avail relevant local services easily ! These premises have huge working areas in several hundreds of acres and you are allowed to move only in assigned area mentioned in your permission pass !

So, when you take this care specific to that area's severity norms , rest of the work is as per your normal routine . The local team is available with you for any type of help required in the middle ! It's their job also but they wish you must complete your scope of work to best possible capability !

12.5 Research and Development Worksites :

These are not commercial worksites but these are experimental worksites ! Here new innovation is tested for its performance requirement and suitable design and development changes are done after noting the results of experiments and trails !

By its name , its access is authorized only to key people in the organization who are building new products for business future . So, every communication is highly confidential and the site work is not visible to general workmen and engineers working in the organization ! This is typical R& D protocol and their worksites are trained to keep their experimental finding professional !

What type of inspection is done here ? In very few instances here regular inspectors are deputed . The inspector working on R & D sites has sufficient exposure to earlier development activities and he has contributed in product development by suggesting suitable alternative to existing design ! R & D inspector is also a design expert who is well aware about manufacturing feasibility historically ! ⊛⊛⊛

THRILL 13 : THRILLING CHANGES

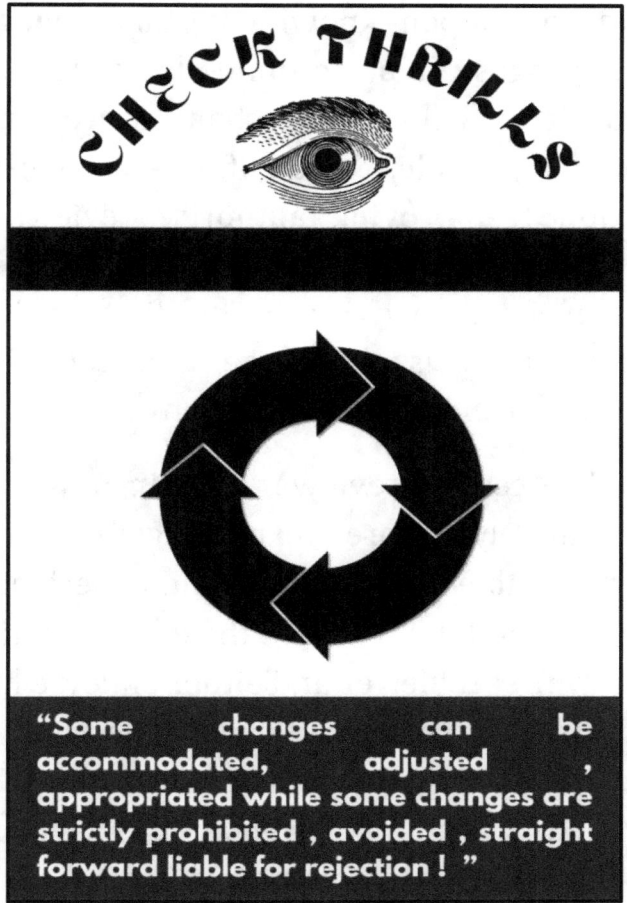

Image Courtesy: Cycle , Pixabay.com

13.1 Introduction :

Engineering activities are human intellect activated processes . As the work is done with human intellect , some types of changes are bound to happen knowingly -unknowingly ! Some changes can be accommodated, adjusted , appropriated while some changes are strictly prohibited , avoided , straight forward liable for rejection ! Approving authority reviews the nature of changes that occurs during every phase of engineering activity ! The first reviewer critically reviews the stage and record his observation and primary decision .

The second reviewer who in most cases comparatively more experienced , more knowledgeable and who has field experience of several functions review those observations made by first reviewer and either endorse his or her decision or suggest alternative decision with which those things will move more swiftly ! They use their overall experience in giving final decision and thus system moves ahead !

Change management is regular industrial practice and there is plan of sending relevant change requisition , approval of requisition after techno commercial analysis, making necessary changes in all relevant documents and sending updated documents with applicable revision or alternation number to responsible heads and then executing those changes with changed material, changed process variable, changed testing, changed identification, whatever may be the type of that particular change!

So, in this discussion, let us see what are different types of thrilling changes and how they are treated techno-commercially !

13.2 Engineering Changes :

These types of changes are initiated predominantly by design department of any organization ! These type of changes consist of change in design data, change in specific dimension to suit exact site conditions, addition of extra connection to existing connections, change in size of present connection to larger or smaller size, change in axial orientation, change in part quantity, change in nomenclature of particular component, changes because of customer induced degree of customization, changes in structural strength calculations and addition of suitable reinforcement through such changes , replacement of earlier part by a customized part , changes in joining and edge preparation details , changes in methods of testing !

These types of changes are circulated to every department after their formal approval by reviewing and approval authority ! The specific role expected from every department is completed with respect to changed document and if these changes are not executed then the inspection engineer remind responsible department and hold further activities till that change is practically implemented !

After physically implementing the proposed changes , formal safety and performance testing is done to ensure the changes in earlier and current performance are practically occurred ! This highlights the fact that proposed change is satisfactorily accepted in the system !

When the new drawing of changed part will be made , the relevant changes will be incorporated in the drawing to avoid any departmental miss outs ! Relevant alteration number will be also changed accordingly . In drawing or bill of material and part catalogue foot notes , the mention of changes incorporated is done systematically and without fail ! In relevant spots in drawing , respective alteration mark is necessary to show for easy identification and further traceability ! These minute details ensure systems full proof ness in dealing with proposed changes !

13.3 Manufacturing Changes :

What will happen if part orientation is required 30 degrees from vertical axis and it is observed on 40 degrees instead ? Need to remove that connection ?

What will happen if part orientation is required 30 degrees from vertical axis and it is observed on 40 degrees instead ? Need to remove that connection ?

Friends , these are some tricky engineering situations which happens during manufacturing incidentally and team need to think on acceptance or rejection of such changes !

In both cases , the wording of changes is exactly same which may confuse any new comer , however in engineering field , when you have to accept the changes , you have to always refer to relevant context and relevant reference !

In some connections , the freedom of orientation can be available from 0 degree to 90 degree or more ! In such cases , even though your connection gets shifted by 10 degrees doesn't alter the overall symmetry and interdependent fitment of parts and hence such type of manufacturing changes are accepted and approved . At the same time , disciplinary instruction is given to responsible work section to follow the approved drawing . In case of any doubts , people are advised to ask immediate supervisor for further technical clarification or relevant upper-level discussion and consultation ! This is done to avoid manufacturing changes which may halt production activities !

In second scenario , the allowable degree tolerance is just 0.5 degree and your connection is rotated by 10 degrees ! So , this is not falling into acceptable range of tolerance and hence this connection has to be safely removed from its current position and you have to fit it as per approved drawing ! This type of manufacturing change is directly liable for rejection !

The allowable tolerance is referred with respect to assembly drawing where relevant part will be fitted on site . If they have to implement the incorrect degree as it is , then designer has to check the feasibility of modification of its dependent connection and see its current manufacturing stage .

If the part is already manufactured , then this change cannot be implemented . If the cost of new dependent connection is less than the rework cost required on main unit , in such case additional dependent connection is prepared inside shop or in 'Quickest approved vendors place' to make this change feasible !

13.4 Material Changes :

Your approved drawing is a compulsion for manufacturing ! You have to obey all its instructions and if you observe any deviation to those requirements , you have to raise a red flag and ask for its clarification from responsible authorities in written and documentary evidence way ! In a big engineering set up , nearly thousand to two thousand small decisions are taken by team while working in such organization and hence there is no meaning of any verbal clearance or clarification ! You have to take approvals in written and systematic form or governing documents with mention of latest document revision number . If this system is not followed , then the decisions are liable to fail or misunderstand !

Material changes are initiated by any department but they are authorized by quality department ! If you don't have given material available with you, then people search for alternative material which has equivalent or equal performance characteristics although there are major price differences !

Here first design department make sure the technical feasibility of alternate material and quality department ensures the safety concerns and additional testing required in new material specification ! At the same time , whether the code of construction allows new material is also checked by quality department be referring to approved material list under different parts listing !

Other material changes are about change in quantity , change in size , change is shape which are easy to implement . In some cases , some additional material heat treatments are added or excluded according to its technical need . Such changes need to be critically monitored at respective stage and suitable system lock need to be in place so that this change may not miss in the middle of various urgency !

Material changes need to be included in Engineering change note , bill of material , part catalogue , site instructions , approved drawings , relevant process map before ensuring its formal implementation ! If any link does not receive this technical communication on time , then there are chances of major deviations ! To avoid such mishaps , central change management function informs such changes through relevant group communication or such changes can be made visible in approved public platform and notice boards inside the organization ! In many cases, a permanent ink stamping is done to mark number of pre-release changes applicable to particular drawing !

13.5 Site Induced Changes :

These types of changes are also known as servicing changes and they are initiated by site servicing department ! When your job is manufactured in shop, the parallel construction of its residential site is going on at customers premises . Every industry has to find a golden mean between readiness of product and readiness of site within allowable tolerance limit, let's say , within a three day to seven days waiting period ! In this period , site people can do other type of finishing work before the product reaches site !

So , if there are any irreversible changes happened during residential site preparation ,it is transferred to design through sales & marketing . It is implemented in drawing and modified part later meets to site assembly truly! This is the value of timely feedback ! ✪✪✪

THRILL 14 : THRILLING RECEIPTS

Image Courtesy: Packing , Pixabay.com

14.1 Introduction :

A big truck is entered in your bought out receipt section which contains material worth 50 lakhs, it is arrived at 4.00 am in the morning! Material required to shop within next two days!

A tempo carrying material worth 2 lakhs is entered in your vendor supplied receipt section as a part of his third trip in a day! Time : 3.00 pm in the noon! Material required to shop on next day's second shift starting from 3.00 pm!

A container carrying imported material worth 5 crore is waiting in your export receipt section! Time : 10.33 pm in the night! Shop is following up for this material since last five days and the major work is on hold!

A stationary trader is arrived in receipt section by taking material worth 5000 Rs 'by hand!' Time : 1.30 pm in the noon! Newly joined team of engineers filled the indent of same before a week!

Friends, this is what the thrilling receipts are! In all these situations, you have to accept the material as per its laid down material and supply specification and forward the purchase invoice to further approving authorities for necessary financial clearance ! The most thrilling part of these receipts is material get in warded either before time , on time or after several follow ups and fire fighting ! This is because , if material is not available on time , people try to cope up the delay till parallel working can be done , but when this time slot is over , then job get on hold because of short supply of material ! Hence , a material manager , production manager and quality manager along with design manager has to work in unison to review the urgent and less urgent material requirement and hence plan the purchase orders accordingly to meet the monthly material demand without fail ! Remember , the time lost in material waiting is not reproducible and it directly account to business losses ! Hence , everyone needs to be pro-active in material planning and its timely availability as per required specifications and ordered quantity !

To facilitate the receipt of material , receipt inspectors can work in general shift and their assistant can work in first and second shift , if required in third shift ! Being a costly affair , certain type of material is accepted only in day time because of their peculiar nature , such as some type of furnace oil , gas cylinders , highly inflammable industrial material !

In this discussion , let us see , how many different types of material is in warded in any organization !

14.2 Consumable & Non -consumable :

Consumable material is required regularly and you need to approve more than one supplier for this and you have to allocate equal ordering at each supplier so that you can get the required material without any delay ! Basically, a safety stock is maintained for such type of material !

Non -consumable material is basically used as supporting material for production activities . If this material is not available ,then production activities halts ! This can be any type of process tools , machines , accessories which are required to speed fast production activities ! Compared to consumable , their requirement is low and specific suppliers are finalized to ensure regular long-term supply !

14.3 Bought out and Semi finished :

Bought out receipts are supplied by its special manufacturers . These parts cannot be produced in manufacturing organization and hence they need to procure from a leading manufacturer of that expertise ! In an automobile plant , engine can be bought out, music system can be bought out , headlights and tail lights can be bought outs !

In semi-finished receipts , few parts are supplied to approved vendor for its processing . At the vendor end , the required process is executed and it is tested by its approving organization's quality representative usually known as supplier QA Engineer !

When this receipt is supplied to shop , the balance operations are completed in shop after due inspection of supplied parts quality specifications . A semi-finished casting in which rough machining is done and detailed machining is balance can be called as semi-finished receipts ! This is done to save valuable time of shops and to develop suppliers' strength to work along with big manufacturers !

14.4 Short Supplied and Replacement Receipts :

These types of receipts are known as repeat receipts because here you have to refer old purchase order for its detailed verification . If you try to refer these receipts without reference of earlier purchase order , you will not able to clear the bill of this receipt because short supply receipts are either processed partially or they are kept on hold till 100 % ordered supply is done by the supplier . Its clear systematic indication is available in every value chain present in the system !

Same logic is applicable for replacement receipt . The original invoice is kept separately in replacement invoice file and when the material will be in warded , the respective replacement invoice along with available documents will be reviewed ! Every receipt function has to file daily receipt invoices systematically so at the end of every month the dash board can be matched with respect to material ordered , material in warded and material in waiting ! You have to tally simple mathematical formula to ensure correct operations at receipt function . The formula is :

Material Ordered = Material In warded + Material in waiting

Case 1 :

e.g. 5 Cr (Ordered) = 4.9 Cr (In warded) + 0.1 Cr / 10 Lakhs material in waiting

Material Procurement Efficiency = 4.9/ 5 = 0.98 , In percentage = 98 %

Case 2 :

10 Cr (Ordered) = 7 Cr (In warded) + 3 Cr in waiting

Material procurement Efficiency = 7/10 = 0.7 , In percentage = 70 %

Case 3 :

50 Lakhs Ordered = 48 Lakhs (In warded) + 5 Lakhs (In waiting – 2+3)

Here account value of inward material is just 50 Lakhs , the balance 5 Lakhs material contain 2 Lakh cash value and 3 lakhs credit purchase , which will be invoiced later on receipt , this is done to take advantage of rate changes in the market !

So , here

Material Procurement Efficiency = 48/ 50 = 0.98 , In percentage = 96 %

Waiting 5 Lakhs material is to be taken into account in next month !

So , it can be easily said that material receipt performance decided the speed of in process work . A plant which receives material with material procurement efficiency like 98 & 96 % is going to produce parts much faster than a shop which receives material with 70 % efficiency !

For this reason , there are typical receipt clearance time allocated for executives , within that time material has to be inspected , rejected and recorded in receipt store. Whatever may be the decision of receipt either acceptance or rejection, it has to be communicated in receipt system within allocated time so that material buyers can initiate necessary corrective and preventive plan with suppliers !

14.5 Domestic and Imported Receipts :

Material which is in warded from your national boundaries is known as domestic receipts. They are easily procurable if sufficient stock is maintained by its regular supplier ! Buyer has to discuss the quantity, per unit price , any commercial discounts or offers and they have to place purchase order mentioning terms of supply clearly !

Imported material is traded on international platform and basically, they can be easily seen on verified official websites. Here all contact details are given with which you have to contact the supplier and make deal of your purchase . Before starting any imported purchase decision, you have to find out the applicable taxes and duties, any surcharge or cess, any international security amount, the current forex rates, statutory permissions or licenses required to procure protected material. When you comply to all import purchase norms, your order is placed and it is dispatched either through air way or by sea way ! The scope of supply has to be clearly defined in your imported purchase order as per import-export terms !

14.6 Stationary & Food related daily receipts

If the organization host internal canteen facility, then the relevant material is procured by canteen in charge as per their service contract. This material can be procured on daily basis or weekly basis depending upon the type of material used in canteen operations.

Same thing is for stationary or administration related receipts. This material is required for on time fulfillment of official documentation . This material can contain stationary items, printer ink, blank pages, drawing sheets, pen, stickers, staplers, files ! At the receipt end, they have to refer to provided invoice and hand over to respective function !

THRILL 15 : THRILLING DISPATCHES

Image Courtesy: Delivery , Pixabay.com

15.1 Introduction :

Scene 1 :

The month of year is April and your team is planning the first dispatch of the season . Incidentally ,it is first export of the year also ! The things are ready in all respect and you are comfortably meeting the planned shipment schedule one or two days earlier ! Team is happy with getting excellent start of the year by delivering invoice worth 5 crore !

Scene 2 :

The month of the year is November , typical winter vacation season ! In this third quarter of the year , your team is delivering 10 back logs of second quarter and this month's 40 jobs ! Overall , there is plan of 50 dispatches in which 15 are export orders and 35 are domestic orders ! The stage is set in such a way that in the last week of the month , you have to dispatch minimum 8 jobs per day to ensure you meet the target of 50 jobs precisely and qualified for your six-monthly performance incentive ! Your dispatch department has provided detailed schedule of shipment and priority of dispatches

! With this schedule , you full fill all orders by working till daily target is met ! Your team efforts are phenomenal and, in the end, you achieve your dispatch goal !

Scene 3 :

The month of the year is march and, in this month, you have to complete all orders of annual business plan ! Your annual business plan is made for 200 Cr turnover and your orders in hand consist of value of 400 Cr ! You are in a favorable position of order booking for next two year and hence you have to complete this year's last quarters target of 60 Cr ! Mathematically , every quarter is supposed to full fill the goal of 50 Cr but in practical conditions , last quarter get the target of 10 Cr additional because of quarter wise order appropriation !

So , while planning the time table of available 26 working days and 4 holidays , you have declared over time schedule and those who are willing to attend the work in this month will get the overtime payment as per their total duty hours ! The daily invoicing target is 2 Cr and this has to be meet swiftly ! All material is procured for last quarter , all manpower is available , war footing management decision system is active and the team has to just deliver the target swiftly ! With proper planning and precise follow up , it is observed as ,you have achieved your target one day advance ! The team was celebrating the annual business plan achievement on 31st march!

So , what do you understand from these scenes and situations ? Answer is very very simple ! The final figure of completed dispatches is the result of annual team contribution . If the team is successful in making annual target feasible , then that confidence give birth to next year's annual business planning and resource addition ! If your next year's orders in hands are more than last year and if you have achieved target of last year swiftly , to achieve the target of next year's 25% rise in annual business plan , the management will increase the resource allocation and this will boost the process of employment generation ! The result of these decisions will also improve business growth avenues !

The condition of quarterly and annual dispatches depicts exact picture of work -culture in any organization ! With this introduction, let us consider the types of dispatches and their importance in industrial goodwill generation.

15.2 Domestic & Export Dispatches :

Jobs which are dispatched within your national boundaries are known as domestic dispatches . Jobs which are exported outside your national boundaries are known as export dispatches !

Every order processing has certain flow of information regarding job completion and its payment schedule . Usually , orders are book with 15-20 % advance payment , 50-60% interim stage completion payment and 35- 20 % final payment before dispatch or withing agreed credit period of 30 days or more as per terms of contract !

Once the product is dispatched from manufacturing site and it is received safely and securely at customer site , the documented formality of balance payment completion starts and product is installed in customers location ! Customer become owner of that product !

In Export orders , along with formal documentation , statutory permissions from local authorities are also required before using that product . If the products are already approved for international supply by certain type of approval , then they can be directly used by looking at its international approval tag in product description !

You prepared same product but export product gives you more profit ,because production of this product is not possible in receivers country or the cost of manufacturing in that country may be far higher than yours ! Only in these two critical conditions export orders are given to efficient manufacturer !

15.3 Regular versus On-Hold dispatches :

Regular dispatches refer to orders of repeat customer or customer who exactly follow your order processing path ! This means , in regular dispatches , as soon as your stage get complete , you get the stage completion payment within 1-2 days , you get advance as soon as purchase order is received , you get final balance payment before dispatch of the job or within 30 days after dispatch ! These dispatches are known as regular dispatches where both manufacturer and purchaser follow the terms of contract precisely and within acceptable tolerance !

On-Hold dispatches are critical dispatches which eat your inventory space till uncertain reasons . Off course the responsible party has to pay the extra charges of delayed product lifting or delayed product dispatch !

Delay can happen from manufacturers and purchase depending upon the business scene at their end ! Any type of business uncertainty and occurrence of unpleasant event causes hold on processed orders in between the processing and that job is kept in separate section for later processing ! When the things at default end sort out , the official communication is received to process the job with payment of balance charges and on completion of the order , the dispatch is done ! These dispatches can be delayed from 1 month to 11 months' time period . All late delivery charges and penalties are mutually acceptable as per purchase contract ! For any unlawful deviation , both parties can approach respective jurisdiction for expected justice !

So , your ability to handle the waiting inventory is your business strength and financial acumen ! Bigger is the manufacturer , higher are the waiting days till the matter get resolved ! Similarly , bigger is the purchaser , faster will be dispatches to avoid huge late delivery penalties ! Big customers need jobs or products in quickest cycle time and for this they are ready to pay premium pricing which is 20-30% extra than your normal pricing !

15.4 Product versus project dispatches :

Product based dispatches involves least amount of customization and higher amount of standardization ! Order processing of product type of dispatches follow a pre decided standard path and there is no change is order unless and until specifically mentioned in the document!

So, one can plan product realization even before they receive typical orders from customer ! Product type orders doesn't wait for customer approval but they are worked based on typical market trends and forecasts ! Suppose , you come to know that seasonal demand for your product is 1,00,000 units ! You immediately start your product manufacturing with standard design of product ! At the end of quarter , you produce 1,20,000 units as whole and in actual market you are able to sell nearly 1,18,000 units of that product ! It means , trend and forecast are exceeded by your production capacity ! The remaining 2000 units are procured by your friendly business associate with business-to-business transaction and for this deal , he paid you 5% extra charges as the demand for this product at his end is also high ! So , this is about product type of dispatches !

In project type of dispatches , there is low standardization and higher customization ! Because of higher amount of customization , you have to strategically think about how the work can be planned so that orders are fulfilled in respective phases of dispatches ! Project orders need scheduled dispatches , not early ,not later , just on time ! When one phase of project completes , then work of next phase starts . In some projects , first and second phases are running parallelly and till these phases get completed , the work of third and fourth phase cannot be started. So, accordingly project manufacturer has to supply his scope of dispatches for that project ! Number of manufacturers may be involved in typical project environment and hence fair co-ordination is must to maintain the phase wise dispatch plan ! A project gets completed only when all its phases are delivered on time and payment get fulfilled as per your agreed payment terms !

15.5 Low value versus high value dispatches :

This is nothing but strategic arrangement done in dispatch planning to ensure fund liquidity for that month ! Low value dispatches are completed in less time and one can expect their payment within 1-7 days ! High value dispatches take longer time to complete and their payment can be received within 30 days or more as per your terms of contract ! This type of classification also helps to provide balanced load availability in shop floors as well as to plan material purchase where diverse funds can be invested for material procurement ! ✵✵✵

THRILL 16 : THRILLING CALLS

Image Courtesy: Meeting , Pixabay.com

16.1 Introduction :

Engineering inspections are calling based activity . Usually, the manufacturers complete manufacturing of typical stage and then express his readiness of that stage as per applicable specification . On this information , inspector visit that place and observe the accuracy of that stage and communicate the decision of acceptance or rejection !

There can be few conflicting situations about rejection of particular stage . Here inspectors have to persuade for that observation with respect to drawing and make sure that other view is also heard and clarified !

If you go for inspection without formal inspection call , you may find the stage is not yet ready , part fitment is going on, few cleaning aspects are balance , safety barricades are not placed in inspection bay for restricted access during inspection , there can be fitment of other fixtures necessary for stage completion and hence there will be less space available for movement ! Hence inspection is always done after formal inspection call when stage is complete in all respect and safety arrangement is done to carry out inspection ! In this discussion , we are spreading light on different types of thrilling inspection calls which are raised in different manufacturing conditions !

16.2 Regular Inspection Call :

In every shop , there is typical work loading for one shift . As per anticipated readiness of stage , the tentative readiness of number of stages is shared at the start of the day in shop communication book and when these stages get ready , they are inspected as per regular practice ! The main feature of these types of inspection call is both shop supervisors and shop inspectors are fully aware about the status of the stages and they have to just go and witness the stage inside their shop premises !

At the end of the day , the number of stages checked and number of stages found satisfactory is shared and recorded in shop communication book to find out the rework percentage in inspected stages ! Less the rework , better is the efficiency in the shop and cycle time will be lesser !

16.3 Supplier End Inspection Calls :

This inspection is carried out at exterior premises of your approved supplier . Here you visit the place as your organization's representative . Before raising this type of call , your internal purchaser discusses the readiness of stage with supplier and suggest him the required plan of readiness . When supplier confirms stage readiness with purchaser , he is asked to send an e-mail for that inspection call ! Once the call is received , the inspector arranges the inspection visit and check that stage !

This visit is followed by detailed inspection report in which photographs of stage checked are attached and inspection observations along with purchase order details , approved drawing number , reference procedure number are entered before accepting or rejecting that stage !

Final acceptance of the part is done based on inspection release note signed by supplier inspection engineer at supplier's end ! This inspection report is referred during receipt of material in shop premises or site premises for which it is written & mentioned specifically !

16.4 Problem Solving Inspection Calls :

When you are working on first of a kind of job in your workshop or at your vendors premises , there are chances of few inspection calls specifically meant for problem solving !

For new products , drawings are designed by designer by using his or her best of subject expertise . But when that drawing reaches shop after basic drawing review , some manufacturing related hurdles are observed ! These hurdles need to be resolved and necessary drawing changes need to be done as per shop or vendor's feedback ! Such type of calls are known as problem solving inspection calls !

Once a field decision is taken , the responsibility of inspector lies with its documentation inclusion and wide spread communication to all concerns so that they can note the changes in earlier assembly ! Usually, a typical model is loaded at 4-5 vendors , hence if new problem is noted at first vendor , same is communicated to other four so that they will not waste the time in set up rework and can go ahead as per new decision communicated in this regard ! Such calls can be attended by other cross functional engineers to facilitate those changes across all levels of manufacturing !

16.5 Third Party Inspection Calls :

Third party inspection is an inspection contract service provided by registered inspection professionals for their customers . Here customer issue a purchase order for supply of inspection services on call basis at designated manufacturing location on behalf of them !

Third party inspectors have to visit manufacturing location , carry out inspection and suggest the decision of acceptance or rejection to customer with the help of inspection report ! Three copies of inspection report are generated , one is made for inspector, one is for vendor where inspection is carried out and third is to customer who placed the inspection purchase order !

Customer who issued the purchase order may see the status of job in final inspection stage before dispatch of the material ! Here third-party inspectors provide their excellent job knowledge during inspection to improve major quality aspects of job ! Third party inspectors are known for improving physical and document accuracy of the job !

16.6 Site Inspection Calls :

These type of inspection calls are generally raised by your customer during service phase of your product ! They can be considered as customer feedback , customer query or customer complaint depending upon the type of inspection call !

You are manufacturer of your product and customer is its user ! You are aware about detailed manufacturing process of that job while customer knows how to use your product ! If during use ,some mistakes happened and product starts malfunctioning , customer to avoid further damage , calls manufacturer's representative on which service engineer is sent to that work site ! When service engineer notes typical problem related to shop inspection , they communicate the matter to inspection department and they do visit to carry out the typical correction required in job at site !

Here field people are not trained to carry out the rework on main job and hence you need specific people from shop who can do that job . You have to monitor that rework and when it gets complete , you have to witness performance testing and then certify the product for its total safety and expected performance . You have to show particular trials after that rework completion so customer can accept that job !

16.7 Urgency Based Inspection Call :

During firefighting hours , there are many jobs running on and there is huge requirement of urgent inspection . Urgent inspection means not signing the inspection report because of urgency but making inspector available for inspection as soon as stage is completed !

Few products are manufactured at remote facility and inspector from purchasing organization is working in different work shops . When such type of urgent inspection is required , one inspector form purchasing organization is permanently deputed to vendors site so that they can show the stage as soon as it gets ready !

Here the assigned inspector has sole duty to get that particular work done within service hours and share its status to all its stakeholders through formal inspection report ! Because of this type of inspection , proper expedition of order happens and inspector act as common decision maker between purchasing organization and vendor or supplier !

16.8 Statutory Inspection Call :

These type of inspection calls are legal requirement of your product and service profile . Unless and until you follow these inspection requirements , your product or services will not be certified by statutory inspection authorities and hence till that certification , your product or service will be known as ' Unregistered product ' or ' Unauthorized service ' ! To avoid this classification , you have to apply for approval with regulatory form and its prescribed fees , you have to confirm scheduled appointment of statutory inspectors and has to keep your work ready in all respect ! On their inspection visit , they will check the things and they will authorize or reject the stage under checking !

In case of approval, you can go ahead for further processing, in case of rejection, you have to take next appointment by filling prescribed fees and making error rectification !

Such type of inspections are formal requirement of your product facilitation profile and you have to adhere to definite timelines so that there will be no default from your side for which you may have to explain in writing and in some cases, you may have to fill late fees of inspection !

16.9 Customer Inspection Call :

Where customer has loaded their order first time or where the job delivery is important and its quality specifications are extremely stringent, a customer inspection is put into place and there is regular interaction with vendors regarding the status of job ! Few jobs are manufactured for year long time. In this tenure customer keep visiting the workshops after every month or after every two months to give approval for completed products. Because of customer approval site work become easy ! ⊛

THRILL 17 : THRILLING LOSSES

Image Courtesy: Financial , Pixabay.com

17.1 Introduction :

The ultimate aim of business is assumed as making continuous profit and use that profit to grow business further ,but real meaning of business is to provide profit making solution for your customers !

This is because , in practical scenario , if your product is giving more profit to your customer than rest of your competitor or none of other competitor , then he or she is going to procure your product for long time and repetitively and this is going to give you continuous profit over longer period of time .

Same time the goodwill generated from your current customer is going to spread the success story to other customers and they are also going to associate with you ! This is how profitable businesses work ; they always focus on profit of their customers before filling their own kitty of profit !

If business is so simple then why businesses suffer losses and thrilling losses in particular ? In this discussion , we are spreading light on thrilling business losses and their effect on businesses !

17.2 Technical Losses :

Many time industries launch new products , some get sold easily but some wait for their customers for long time ! In some cases , this waiting is extended till one or two years after product launch , but this product cannot attract sufficient number of customers with which the continuous production of this product can be run in respective shop floors and with which business growth momentum can be achieved !

The whole efforts made by trend and forecasting team , design team , manufacturing team goes in vain and team has to again think on exact market demand and its sufficient order generation potential ! If you are running a business , this fundamental marketing knowledge , skill and decision-making ability is very very important . If you lack this ability , then the whole plan is going to affect in the end and you are not going to achieve required order booking and its intended sales to generate required amount of profit !

The main reason of such losses is product misses' key features which are available in other producers' products , secondly the product can be comparatively expensive than other competitors , thirdly people don't like to go for other products till they really find better option than current working product ! This all result into technical losses !

In some cases , critical environment & personal safety related technical issues are observed in the product offering , which also cause reduction in product demand and hence possible business losses ! Organizations has to call back the supplied products or they have to repair and replace them at free of cost , if serious technical issues are noted post sales !

17.3 Management – Union induced conflict related losses :

Most of the traditional manufacturing organizations has regular management- union industrial relations ! In software industries , there is fair professional relationship between management and recruited staff ! Profitable businesses always have a smooth management – union relationship ! In this type of mutually rewarding relationship , management express the current business scene and asking rate of order completion while union representatives express the tools – tackles and relevant facilities along with fair compensation rise for their mental- physical efforts ! After fair discussion round , both agree to a practical proposal and work according to that proposal ! At the end of every month the expected and observed performance is analyzed from both side and corrective measures are shared to accelerate the performance to next higher level !

At the end of the year ,annual business performance is analyzed and whatever was promised during management -union meet is provided as per terms of contract ! In this case , both parties have achieved their goal and it is generally called as a mutually ' win-win ' contract ! Because of such contracts , many employee and workmen welfare schemes can be initiated with initiation and participation of union and management and employee keep receiving the fruits of their dedication and commitment to their work profile !

In a conflicting management – union relation , there is constant disagreement on various operational and financial grounds and this disagreement actually reduces required rate of production per month ! When the monthly production goal is not achieved up to bare minimum profit generation threshold , business cannot afford to making payment to their suppliers and thus experience financial burden !

If this type of situation continues for long time , then the production goal goes on increasing , when customers observe their orders are getting delayed beyond their agreed tolerance limit , they may induce performance penalty , in some unpleasant situation , they may cancel or hold the order ! All such effects affect your business adversely and at the end of that year , your business record net loss for that year !

If management is sensitive and responsible enough to take care of business , it analyzes the people or subjects which are responsible for these losses and they find out practical solution to remove those system hurdles fairly so that their business reputation remains intact ! Then again, the next years running performance is observed and if system get improves, business start making profit again and if system still behaves in conflicting mode, legal proceedings, customer dissatisfaction, then that business keep making losses for that year also !

Such type of business survives till it has necessary financial liquidity and other business assets owned in good time of business ! If suitable conflict resolution action is not taken by management or if union not supported to managements repeat work condition improvement proposals, then such organizations reaches to judicial procedures and the main aim of product generation to serve customers remain unattended ! Such organizations are later get closed permanently !

17.4 Industrial Revolution related losses :

Change is only permanent thing in business and organizations which adopt to agile management techniques absorb industrial changes swiftly and make their internal and external system aligned to remain in the business till another cycle of industrial revolution !

You have prepared 1000 cars and they are available for purchase , but in last 20 days , your competitor has launched a new car which has artificial intelligence related modern communication system , or the competitor is offering 7 seats in the price of 5 seats model , this technological revolution is going to compete with other competitors also ! Your customers are going to change over the period of few hours and thus before this situation occurs , you also need to develop similar product in the market within same competitive range ! This is what survival in business is !

So , there are many brands which changes themselves with respect to industrial revolution and keep themselves updated about what others are developing ! This study is very much critical and is the basic of avoiding sudden business losses ! By the way how many people still use simple mobile when large variety of smart phones are available ? How many people build houses of mud and clay when the RCC technology is gone so much ahead that they can construct tower more than 100 floors ? This is what industrial revolution related losses ! The small or big businesses who are not able to adopt these changes , keep making business losses because of lack of demand for their product and this causes no supply as there is no production ! This demand-supply vicious circle is extremely risky for any business and hence organizations have to continuously upgrade themselves according to latest technological inventions which are productive and economical than your current business practices !

17.5 Natural Calamities, Geo politics and Industrial Safety related losses :

Business existence is the prime requirement of business profit ! To make this profit , there has to be constant product demand , suitable policy environment and absence of natural disturbances to workflow !

Natural calamities like earthquakes , floods , landslides may affect your business existence adversely . You may need to halt your business operations for these reasons which may eat your productive time . Because of these reasons ,business incur business losses !

Presence of geo-political issues may affect your business adversely . Rise in taxes and duties , other overheads , legal provision for employee welfare , working hours and weekly off changes , government intervention in material purchase and labor -staff recruitment rules may reduce your current profit margin and thus if right equation in not adjusted , your business makes losses !

Industrial safety is of paramount importance and if it is neglected , then there are chances of unpleasant incidence one day ! Such incidence will cause damage to your property and people's lives to great extent and rebuilding efforts in current condition will be highly expensive which can cause further burden on your reserves and surplus ! Till the time ,your production is under halt , your management is fighting legal battles , this loss remains there! ⊛

THRILL 18 : THRILLING HEAT TREATMENTS

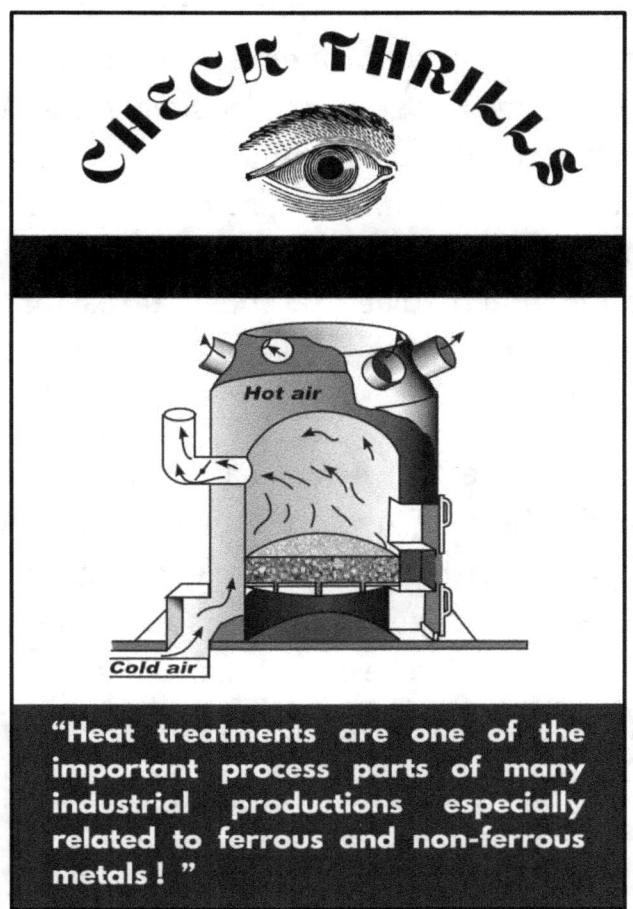

Image Courtesy: Heating , Pixabay.com

18.1 Introduction :

Heat treatments are one of the important process parts of many industrial productions especially related to ferrous and non-ferrous metals ! Heat treatments are done to improve existing physical properties of metals like hardness, grain size , corrosion resistance , tensile strength enhancement or reduction to suitable desired limit , phase transformation stabilization and to get a homogeneous uniform metallic microstructure post stressful cold working or hot working to remove residual stresses !

The effect of heat treatment is observed with the help of heat treatment graph which consist of plotting a scale of treatment on x axis of time and y axis of temperature . The swift movement of process is recorded on process graph !

When the heat treatment is completed , metallic job is either furnace cooled , air cooled or quenched in water to get the desired microstructure and mechanical properties sharply ! While technical success of heat treatment lies with study of present and required microstructure as per respective phase diagram of metals , its mechanical properties success depends upon the accuracy of operations which involve raising the temperature based on thickness calculation , lowering the temperature based on soaking time calculation and soaking time based on required transformation temperature below or above the critical temperatures of that metallic phase diagram !

In this discussion , we are going to see some of the thrilling heat treatments and what type of engineering inspection is required when job is removed from heat treatment furnace !

18.2 Pre heating :

In this type of material heat treatment , part is loaded in furnace or near the local heating source and hitted uniformly so that the moisture content is removed and material become dry and free from any oil like contamination ! If pre heating is not done , the presence of moisture and oil particles create pinhole porosity and blowholes which detrimentally affect structural integrity of that part !

18.3 InterPass Heating :

When you are carrying out processes like welding , to initiate slow weld pool solidification , you have to provide interpass weld heating . Because of this additional heating , weld puddle solidifies slowly and thus it avoid hard microstructure and less heat affected zone . Lowering of HAZ is beneficial for the weldment while widening of HAZ is prone to intergranular weld cracking ! Depending upon the number of weld passes , intermediate heating during interpass is done ! This temperature is different for different material !

18.4 Post Heating :

This is again done with the help of local heating source so that final weld solidifies slowly and uniform weld microstructure is achieved and same time temperature difference between parent material and weld metal will be minimal

! If the post heating is not done , surface cracking in the form hydrogen induced cold cracking of micro cracks starts and if such part is loaded under stressful conditions , such microcracks and fissures tend to elongate and get converted into fully developed deep crack which are bound to fail in brittle way !

18.5 Post Weld Heat Treatment :

Here the part under heat treatment is either locally heated to post weld heat treatment range and soaked at that temperature depending upon part thickness . After soaking , material is brought to room temperature by slow cooling in air ! The main purpose of this heat treatment is to remove the residual stresses induced in material during welding and set up processing !

Post weld heat treatment requirement is not required for some material while for some material it is required when joint thickness increases beyond certain range as per design code of manufacturing ! Along with job , you have to heat treat test plates to check properties of parent metal and weldment after heat treatment

! There is notable color change after post weld heat treatment !

18.6 Solution Annealing :

This heat treatment is done for stainless steels . Stainless steel of certain grades is heated to solution annealing temperature between 1040 degree Celsius as per thickness of the steel and then immediately quenched in solution bath to bring the carbide into solution !Chromium carbide is a strong and hard phase !

Stainless steel is known for its high corrosion resistance which is because of protective chromium oxide film which forms after reaction with air !

In corrosive atmospheres , the chromium carbide presents in the microstructure protect the grain boundaries and thus metal remain free form corrosion attack !

If this treatment is not done , grain boundaries get attacked because of corrosive environment and this corrosion give rise to metals in service failure !

18.7 Hardening :

Hardening treatment is done to increase the hardness of the metal ! The material is heated to hardening range and then suddenly quenched in oil , water or resin ! When the material hardness is checked , rise in hardness is observed ! Hardening is done for parts which requires good amount of wear resistance !

18.8 Annealing :

When you want to have fine grain microstructure for sharp properties , the material is heated to annealing range , soaked for certain hour and then cooled in furnace . Furnace cooling leads to slower cooling and because of which fine grained microstructure with great tensile strength and ductility is achieved !

18.9 Normalizing :

The basic process of annealing and normalizing is same , except normalizing is done for getting coarse grain microstructure ! For normalizing , material is cooled in air , that why it is known as normal cooling or normalization !

18.10 Carburizing :

In industrial hardening when you want to keep core as it is and case to be hard , the carburizing heat treatment is done ! In carburizing , the surface hardness of metallic part is increased by carburizing process in which carbon get added in surface of the metal ! This additional carbon increases hardness of case of the metallic part !

18.11 Nitriding :

Nitriding is also a hardening process in which nitrogen is added in case of the metallic part to improve its hardness !

18.12 Carbonitriding :

This process is the combination of carburizing and nitriding and it is done to enhance the surface properties !

18.13 Tempering :

Usually, many hardening heat treatments are followed by applicable tempering heat treatment in which ductility of parts is improved by heating the metal again to a lower tempering temperature and then cooling it to room temperature !

18.14 Ageing : Age Hardening or Precipitate Hardening :

Few metals lose their strength & hardness after some servicing time ! In ageing heat treatment, this strength reduction is reduced by carrying out a heat treatment known as ageing ! So that they remain strong for longer servicing time ! Here supersaturated solution transforms to precipitate particles that enhances strength !

In Maraging , steel is heated so that martensite – a hard phase is formed and then it is air cooled so that its hardness gets improved !

18.15 Martempering :

In martempering ,steel is heated to upper critical temperature and then quenched in hot oil or molten salt bath kept at tempering temperature around 150-300 degree Celsius ! This heat treatment improves impact energy and fatigue resistance !

18.16 Inert gas protection around weld puddle :

This is not a direct heat treatment but it is part of natural heating which occurs during typical welding process ! A gaseous shield of noble gases like Argon , Helium is provided around molten weld puddle so that foreign particles like dust and oil do not contaminate with weld puddle and thus clean solidification of weld puddle happens !

So , these are typical heat treatments which are done as per jobs requirement . Beside of these heat treatments , if customer insist for new heat treatment as per their code knowledge , then manufacturers have to arrange such customized heat treatment to get properties accurately in the job ! ✱✱✱

THRILL 19 : THRILLING PAINTING

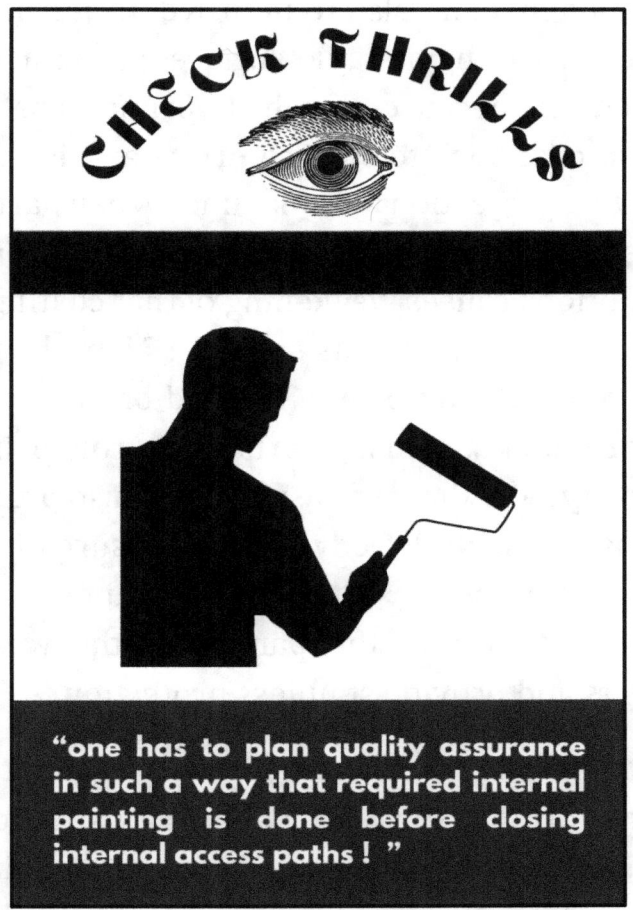

Image Courtesy: Black , Pixabay.com

19.1 Introduction :

Industrial painting is one of the important finishing stages of in process job ! Painting is done to provide protective coating to components and also to improve its aesthetics ! Before painting suitable surface, preparation is necessary because of which primer coat and finish coat stick to surface firmly and it help to provide protection for long time . When parts are fitted in service environment , the weather conditions causes weakening of that coating and it get removed from its surface ! Then the parts are repainted according to need of that service environment . In many parts apart from painting some type of thermal insulation is also attached on parts which directly avoid exposure of paint to environment . Steel or aluminum sheet wrapping around the part faces the weather effects and provide stainless protection !

So , in this discussion , we are going to see some of the thrilling painting inspections , typical painting schemes , number of painting test and risks involved during industrial painting for painters and technicians ! With this detailed sharing ,we will come to know that painting is a critical process in manufacturing any product !

19.2 Internal Painting and lacquers application :

When you have to protect internal parts of products , the internal surface is coated with paint and lacquers . When the parts are to be used for food processing or drug preparation , the internal surface generally consist of stainless steel , aluminum . But when internal surface belongs to material like mild steel , the surface is painted with internal coat of epoxy paint . In some cases, only lacquers are also applied ! The DFT of internal paint is generally less than external coat , this is because it has to face internal controlled environment ! There is access limitation when you are painting internally . Hence one has to plan quality assurance in such a way that required internal painting is done before closing internal access paths ! If the painting remains balance , then you have to remove those access path to get it done . This care is must in any internal inspection and inspectors has to provide internal clearance before closing access paths !

19.3 External Painting :

External painting is important process step and it is carried out in dedicated paint shop . Paint shops are huge open grounds or closed environment with proper exhaust system . One has to take utmost care that whatever fumes and sprays are evolved during painting must exit quickly , if these fumes and spray particles remain in the nearby surrounding , the person present there may inhale those paints and it is not good for your respiratory system . Paints are hazardous and explosive substance and hence all personal protective equipment's along with necessary painting suite to wear by painters and inspection engineer !

Surface preparation of part is carried out with respect to emery paper cleaning or sand blasting . Tiny round particles of sand are sprayed on the metallic surface with pneumatic arrangement . In some cases, shot blasting is also done . When these particles hit metal surface , extremely small pores are created there which create required surface roughness to hold the coating of primer paint ! Primer like red oxide is used on external surface and after complete drying and measuring dry film thickness , next coat of finish coat are applied !

There is typical way of carrying out painting of external part . You have to start from one point and return to same point after making a complete round around that part ! The application of paint is always top to down to provide gravitational force to paint . If you starts from bottom and moves to top , then top paint comes down and it looks like shabby appearance of painting ! Starting from left and moving towards right and reversing the direction of coating is also regular way of painting ! In reverse direction paint coating thickness improved to satisfactory level and unpainted parts get painted correctly !

Spray nozzle has typical sizes and you have to choose required spray nozzle depending upon the space availability and intricate places in parts . Inaccessible parts by spray painting are painted by using hand brush of variable sizes !

Painting is generally done in two types of final finish. One is called as matt finish while another is called as glossy finish ! There is variable protection in both these types of finish. Heated parts are generally observed with matt finish while casings which host heated parts inside them are observed with gloss finish !

When you have to paint bottom most part of any product, you have to lift that part with the help of safe lifting mechanism and by standing just below the part, you have to paint bottom most part ! Paint inspector typically inspect this portion because there are chances that painters may not achieve required dry film thickness in this zone because of access issue !

There are many places in a product where painters have to reach personally to ensure the part is painted. Such places involve back sides of small connections, top most sides of major connections, this may remain unpainted because of its height, padded parts which are covered before painting, there are chances that these covers are not removed before painting ! When carrying our external painting entry and exit doors, access doors, furnace doors to be kept open so that they can be painted internally and externally . Some connections need to be protected near mating surfaces . Such surfaces are covered properly before sending parts for painting . The remaining paint if any on such machines surfaces is cleaned by using thinner and clean clothe !

Hardware is important material for tightening the doors and access path . Hence you have to take care that hardware is not fitted before painting . Because if it is fitted and painted then it becomes difficult to lose that hardware and while losing it removes applied paint . Hence hardware is to be fitted only after complete painting and total drying of paint ! Excess paint below hardware seat to be removed by scale or sharp blade smoothly by rubbing the blade on paint gently ! It ensures firm gripping of hardware's with finish product !

In any type of painting , the paint application to surface is checked with respect to dry film thickness measurement . Painting inspectors are qualified according to NACE (National association of corrosion engineers) Level 1 , Level 2 certifications ! Here they check all technical details required for particular product as per its service environment , necessity of coating , part fitment zone , applicable service temperature and pressure variation , wind pressure and other dynamic and static loading concerns !

Sea worthy painting and sea worthy packing are two important features of specialized paint & packing application ! Here considering the typical marine environment , there is high rate of corrosion when parts are fitted in marine environment or they are transferred from shipping path ! If the paint scheme design is not correct , then that paint spills off while passing through salty sea atmosphere and when it reaches to worksite , shocking experience is faced by customer ! Hence , NACE inspectors ensures the correct painting scheme is designed for part under consideration . There are NACE standards which are referred to provide excellent surface finish and protective coating !

Apart from industrial painting anodic and cathodic protection is provided in process like plating and galvanizing ! In this process , the internal surface is protected and external coating faces the corrosive atmosphere ! The external coating gets consumed with that reaction but still internal surface remain corrosion free !

19.4 Panting of pipelines , valves , heat exposed parts :

This is again a thrilling task . In typical petrochemical environment , there is huge arrangement of pipelines to carry the fluid from one place to another ! Such pipelines are manufactured as per API or related standards and when they are put into service , they are carefully painted to provide atmospheric protection !

Pipelines can be 1 kilometer to several kilometers longer . This length is achieved by joining single pipe of required dimension with number of joints with elbows , tees, reducers ! Here , till joining of part and its satisfactory testing , pipes are not painted . When nondestructive testing and hydraulic or pneumatic testing is satisfactory , then these pipes are painted . For some site joints , pipe is painted partially and area near joint which measures around 200-300 mm is purposely kept unpainted so that there will be easy site joining possible ! If this part gets painted , then during joining with process of welding , this paint contaminates the weld pool and create welding defects !

The valves can be as cast or as forged . Their surface is painted to provide protection from corrosive and oxidizing environment . Valves are directly exposed to surrounding and hence paint schemes for valves need to be designed as per applicable paint standard . Some valves are very small is size while some valves are really huge !

In heat exposed parts like furnace and burners , furnace and access doors , various chambers where heat enters and exist , you have to take special care of coating . Because of frequent thermal shocks , the painting in this area become weak and get removed . Hence paint application in heat exposed area need to be careful ! Apart from this, fair care is required to paint and inspect that painting at height ! ⊛

THRILL 20 : THRILLING PACKING

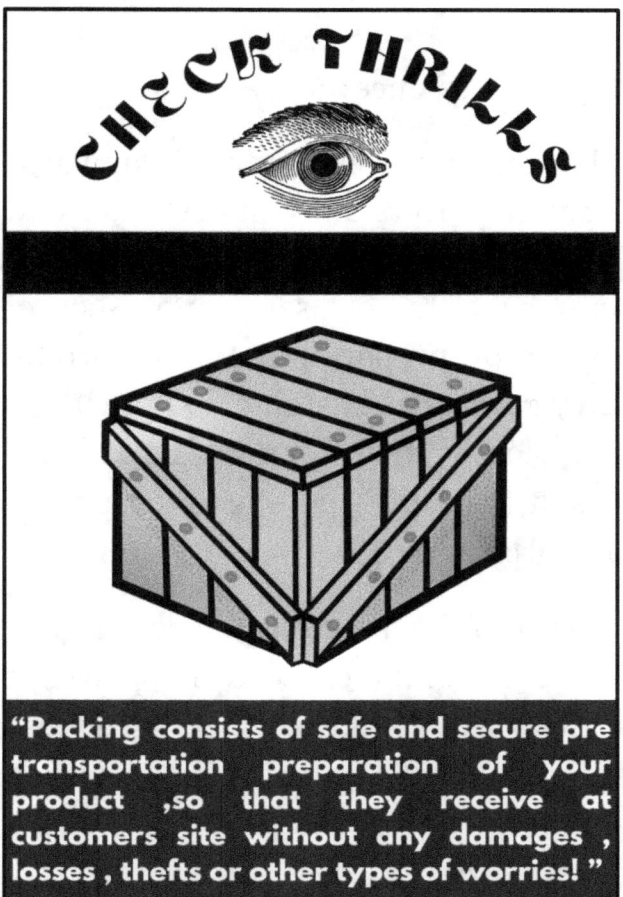

"Packing consists of safe and secure pre transportation preparation of your product, so that they receive at customers site without any damages, losses, thefts or other types of worries!"

Image Courtesy: Box, Pixabay.com

20.1 Introduction :

The first impression of customer and final work in creation or product making is owned by packing process ! Packing and packaging is important process as far as industrial development is concerned !

Packaging consists of total scope of supply from manufacturers side till customer end as a part of specific work contract or work order ! Packing consists of safe and secure pre transportation preparation of your product ,so that they receive at customers site without any damages , losses , thefts or other types of worries that result into any type of short supply as per agreed purchase order !

When parts are transported , they can be kept in closed containers to avoid open exposure to atmosphere ! Also, the parts can be transported on trucks and tempo where jobs are kept under open sky ! Here the role of packing comes into play !

Because of packing , the immediate external part and nearby internal part from where water can enter inside job remain protected. If there is wind, water or heat, the part remains safe till transit and when it reaches site, packing is removed for easy receipts of products!

So, in this discussion, we are spreading light on certain packing requirements, care during packing and losses incurred because of faulty packing!

20.2 Plastic Packing :

Parts which are small to medium in size are packed in heavy duty plastic bags, in some cases bubble plastic sheet wrapping is also done ! There are various sizes of plastic bags and loose plastic sheets are available in the market. The manufacturers hand over the material for packing when dispatch quality clearance is received after due verification of scope of supply and noting all relevant document details which are supplied with dispatch material along with job ! Before plastic packing, the identification tags and their co-relation with scope of supply is tied to respective part so that when the part will be removed from consignment ,one can easily check its quantity and part number . If you miss tagging , then receipt team has to co-relate the same with available documents and in case something received wrong and without tag , it is recorded as major nonconformance to requirement at customer end . If wrong part is supplied which is not tagged properly , the site work may halt till the correct part is supplied again !

However, one can easily remove plastic packing if there is some type of miss out ! After required correction , same part can be repacked with new plastic bag !

20.3 Wooden Packing :

Sea worthy wooden packing is easily available in the market and it is made up from light weight wood ! Before carrying out wooden packing , the part which is packed in plastic bag or bubble wrap is put on wooden base coated with heavy duty plastic from inside to make it water proof ! Then the wooden husk or dry grass of wheat like crops is filled inside that box to avoid shocks and vibrations during transit ! This material is put in excess amount so that part is totally protected from all sides along with top and bottom part ! In some cases , parts are fitted inside thermocol slabs of exterior profile matching shapes !

    This packing is then closed by fitting traditional screws or by using screwing machine which do this work quickly ! The fully packed wooden box is sticked with part details and related tags for easy identification ! On every wooden box , the full details of customer , order number , final weight of that box assembly , caution signs like keep part upright , avoid from falling or breaking , box contain delicate breakable material like glass , box doesn't contain any explosive substance such type of signs to be marked with the help of special ink or paint which remains there on wooden box till part reaches work site ! This process is known as marking of package !

20.4 Heavy duty plastic or Tarpaulin Packing :

Parts which are big in size are covered and packed with heavy duty plastic or extremely large size and tarpaulin blanket ! There are tying rings provided at the bottom part form where one can tie the parts to available lashing with the help of wire or synthetic ropes !

Teflon packing is also used in many parts to avoid leakages in working environment !

20.5 Foam Packing :

When the hot material is flowing through pipelines and it need to be insulated , foam packing is used for such type of insulation . In this case , extremely flexible foam packing is wrapped around pipe and sealed suitably to avoid any loose joints ! This packing remains with those pipes even during servicing and it is not removed till specific inspection of internal arrangement is necessary !

20.6 Cardboard Sheet Packing :

This is most common packing material observed in consumer product packing ! Here part is simply put into well sized card board cover and transported through safe structured racks ! Every rack has distinct place to fit the box perfectly and thus more than 50 products can be fitted in one rack . Such type of multiple racks are kept in container and same is transported to required warehouse !

20.7 Designer Paper Wrapping :

Again, many consumer products are packed inside trending designer paper packing designs which attracts the customer . These types of packing consist of detailed product information , its batch details , its expiry dates , net weight and quantity , advantages of products , tips for safe uses , endorsement and advertising tag lines , brand logo and registration or trademark symbol !

20.8 Special Packing for Food items :

Packaged food items such as wafers , cold drinks , bakery item , chocolates are packed in special packing as per food safety standard . This packing has to take care that material inside it remain safe for use within given expiry date. This case is not applicable for industrial supply like machines and equipment's ,since these are not eatable products ! Hence every product packing requirement is different and dispatch engineers along with product design engineers need to be fairly aware about the typical safety guidelines as applicable for products best performance and safe usage to intended customer !

Because of such food safety regulation, the additives and preservatives are added during manufacturing so that product remain eatable ! So , because of correct process and correct packing , products can be sold to any part of the world without any issues ! Such products are marked with food safety certification along with their food safety license number ! Such things need to be considered before accepting part at customer end !

20.9 Importance of Lashing :

Heavy industrial parts when transported by road , need to be tie with right type of lashing ! Here the part to be transported is kept in the center of load carrier with the help of lifting hook . The part is fitted with lifting hook and lashing hook . Part is tied to vehicle tightly so that it will not fall down in case of extreme turn or speed during transit !

There is arrangement of temporary support which are welded to base plate of product and load carrier . Because of this , part cannot move even though there are speed variation . The remaining safety is provided by proper lashing !

It is observed that heavy transit losses are incurred when parts are not properly lashed and rested on load carrier ! Hence , it is necessary to check whether the parts are properly settled on load carrier !

Along with this mechanical requirement , it is also necessary to have respective insurance for safe or risky transit of products . So ,even though such type of damage may happen , the manufacturers will get relevant compensation for that loss ! If third party final packing -

marking inspection is applicable , then this point is also recorded in report !

20.10 Loose Material with the Job :

This is extremely important and most serious miss out in any dispatch ! These are small to tiny parts and they can be packed in simple paper bags or plastic bags . Because of their small size and quantity ,people may forget to keep them in required place . In total packing , if these small items are missed , then your whole consignment receives at customer end and when customer see the respective loose material is not supplied , the work remains idle ! This loose material has capacity to halt the work even though it is small in size ! Special pins , special loose wire , special hardware , special switches such parts come under loose material . Sometimes spare material is also provided as loose material . Spare parts are used when main part has to be replaced ! So , this type of material to be mentioned specifically either at the start of your packing slip or in the end of the packing slip so that it does not get miss anywhere and it can be traced easily at customer end ! ✳✳✳

THRILL 21 : THRILLING DIMENSIONING

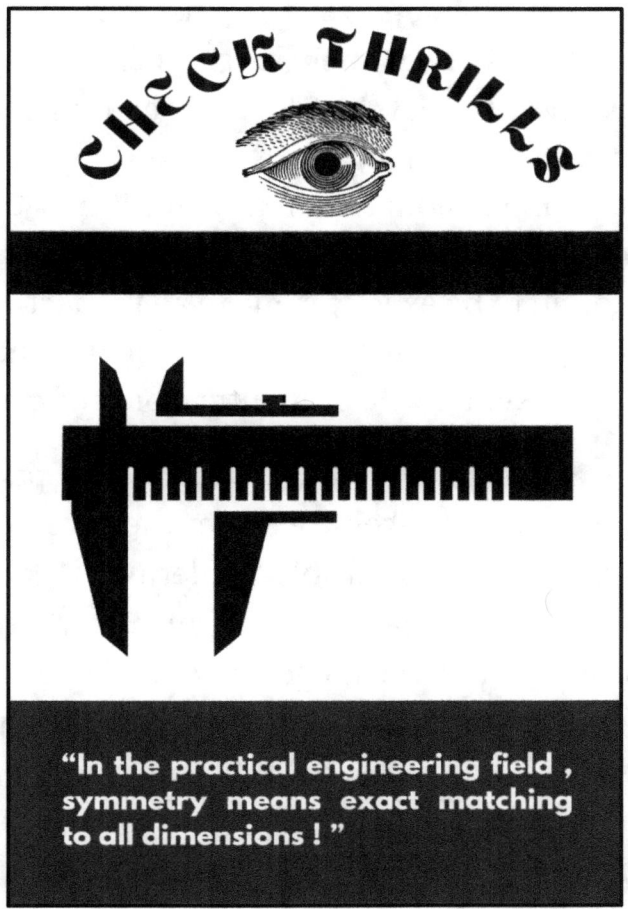

Image Courtesy: Calipers , Pixabay.com

21.1 Introduction :

What does mean by symmetric parts and asymmetric part is question of academic importance ! In the practical engineering field, symmetry means exact matching to all dimensions ! Dimensions include length ,breadth, depth ! This three-dimensional representation of any part with respect to other part is dimensional comparison . When two or more parts compared with one master sample are found to be exactly same in all respect within given tolerance range , we can conclude that the parts are symmetric or identical or uniform !

So, there is no difference in two parts or whatever differences are there , they are present in standard allowable tolerance limit ! Considering human manufacturing , there are extremely minute changes in work done by two different workmen ! This work difference is measured with the help of measurement of dimensions ! If both the jobs show same dimensions , it reflects excellent skill level available with both workmen ! If there are difference in dimension , then one can conclude that there is fair amount of skill gap and one must train other workmen to create most accurate products as per given tolerance limit ! This is what manufacturing excellence is ! Every workman and staff serve to create uniform products without any difference once approved drawing is provided for production !

Second thing, dimensional inspection is objective type of inspection . All inspectors measure the given dimension and they observe dimension available on job ! If the difference in dimensions is within acceptable tolerance , inspector accept the job , if the difference is not in tolerance limit , inspector rejects that job and ask to carry out necessary rework if it is possible , if the dimension is increased , it can be cut to required size , if the dimension is excessively reduced , the applicable joint is provided and the joining connection is tested for its structural integrity ! Best engineering practices avoids joints as far as possible ! But if the joint is unavoidable , then its total joint efficiency , load and stress carrying capacity , strength and ductility is checked with respect to applicable standards ! In this discussion , we are going to see different types of dimensions and their critical check points !

21.2 External Dimensions :

Dimensions which can be easily measured by standing or sitting outside the product are known as external dimensions ! These dimensions typically include total length , total breadth, total height , total outside diameter , outside circumference , outside projection of connection ! Measuring instrument or measuring tape is touched to extreme ends tightly so that there is no gap between tape and measurement surface and the dimension at other end is noted . Some people start dimension measurement by holding tape on 100mm at start ,in such cases , whatever dimension is observed ,that starting reference 100 mm to be subtracted to avoid any type of dimensional error of 100 mm straight ! External dimensions are very very important because many sub-assemblies are fitted inside them and hence these dimensions need to be accurate to avoid any type of mismatching or excessive gap after assembly of other parts ! The overall performance of equipment also affects if major dimensional error is noted in external dimension ! Suppose , final length of tank is decreased by 500 mm in a shell length of 5000 meters , this will reduce the total volume accordingly ! Suppose , total height of equipment is increased by 300 mm in total length of 3000 mm , the mating parts will not match with each other and hence you have to reduce extra height by necessary cutting and rejoining , retesting ! Suppose , the major width is decreased by 200 mm in total required width of 2000 mm , the part which is going to fit inside that enclosure will not go inside as its outside dimension is 1990 mm with 5 mm gap from both end ! In this case opening will become 1800 mm and part to be fitted inside of the enclosure will have external dimension as 1990 ! So , this assembly will get stuck in the start itself ! Hence outside dimensions need to be critically monitored ! The overall use of material is decided by outside diameter . Suppose if you have to construct a cylindrical shell of 1000 mm outside diameter and 20 mm thickness , then required circumference will be (1000- 20) x 3.142 = 3079.16 or 3079 mm sharp after rounding off 0.16 mm ! For calculation of outside circumference mean diameter is taken into consideration and hence one time thickness is reduced from outside diameter ! The required outside circumference is 3142 mm ! So , when the plate is cut and rolled on roller, because of rolling action, the dimension gets expanded to 3142 ! The circumference of cylinder after rolling get increased by 63 mm ! So nearly 2 % increase happens during rolling action !

Now if by mistake, the plate is cut as per outside circumference dimension of 1000 mm, then cut length will be 3142 and length after rolling will be 3142 which will expand by another 63 mm specific to thickness of 20 mm ! In this case, outside circumference will be 3142 +63 = 3205 and observed Outside diameter will be 1020.05 rounding off to 1020 sharp ! So, if mating part to this enclosure is a round tube plate with required outside diameter 955 mm with 2.5 mm root gap for easy insertion inside that cylindrical shell as per cylinders internal diameter (1000 -20 -20) = 960 mm and additional root gap for tube plate insertion 2.5+2.5 = 5mm, the outside diameter of tube plate must be 955mm ! Here observed outside diameter of cylindrical shell is 1020 mm, hence its internal diameter will be (1020-20-20)= 980 mm ! So, the shell ID will be 980 mm and tube plate OD will be 955 mm, hence total gap of 25 mm will be visible in difference of dimensions and if you allocate this difference to top and bottom end , then root gap of 12.5 mm will be observed ,which will be major deviation to design code ! How you will fill this gap ? You need to increase the welding ,which will add more heat input and hence more chances of metal distortion and residual stresses ! Hence you need to resize the shell by cutting excess length of 63 mm and then rerolling to get the required OD 1000 mm , ID 960 mm where tube plate of 955 mm diameter will easily fit !

21.3 Internal Dimensions :

Next important dimensional measurement includes measurement of internal dimension ! Internal dimensions are measured either by going inside the job or placing measuring instruments inside the job ! These dimensions include internal diameter , internal circumference , internal projection ! Here measuring tape's zero end is hold on bottom part and top dimension is taken where tape exactly meets the inside surface ! The tapes are sufficiently strong and hence they do not deflect during measurement ! If internal dimensions cannot be measured because of lack of sufficient natural light , then artificial lighting provision with the help of electric bulb or any type of portable light source to be done before ensuring internal dimensions ! This is the biggest risk in measurement of internal dimension and people may record incorrect dimension because of insufficient lighting !

21.4 Individual and group dimensional inspection :

Dimensions which can be measured by single inspector are known as individual measurable dimensions ! Measuring all dimensions of typical part which has 100 -300 mm maximum dimension and it has any number of part quantity, it can be considered as individually measurable dimension !

In group dimensional measurement ,you alone cannot check all dimensions and you need qualified support to assist in dimensional inspection ! Suppose a part has maximum length 2900 mm or 4500 mm , can you check it alone ? Yes , with 3 meters or 5 meters tape , you can check this length on your own , but if total length is 10 meters ,then what will be the situation ? You need to take qualified assistance to hold the tape at one end on 'zero' and tell the dimension at another end ! Sometime , in few dimensional inspections , when there is urgency of stage completion 2 people can take all internal dimensions and 2 people can take all outside dimensions . Because of this, time of inspection get reduced by minimum 45-50% and stage get easily cleared if everything found confirming to drawing requirements !

21.5 Simple and sophisticated dimensional inspection :

Simple dimensional inspection can be carried out by measuring tape , scale , ruler ! These are straight or circumferential dimensions and easily accessible ! Most of the fabrication dimensions can be measured with the help of simple dimensional inspection !

Sophisticated dimensional inspection consists of measurement with the help of precision measuring equipment's ! These instruments can be vernier caliper, micro gauge, bore gauge, co-ordinate measuring instruments, digital measuring devices ! Here the range of allowable tolerance is very thin and hence accurate recording of all dimensions is necessary ! Tolerance in the range of 10-20 micron is difficult to measure by simple tape but it can be easily measured on co-ordinate measuring digital system !

21.6 Tactical and Technical Dimensions :

Proportionality is an important aspect of every engineering design ! Tactical dimensioning refers to proportionality of dimensional system ! When you are designing a car, then height of car and length of car needs to be proportionate ! When you are designing a house, the height of house and length of house need to be proportionate ! These dimensions are basic dimensions and hence they can be known as tactical dimensions ! Technical dimensions refer to exact dimensions achieved after applying relevant proportionality of tactical dimensions ! Tactical dimension is first raw dimension that comes to engineering mind ! ⊛

THRILL 22 : THRILLING TESTING

Image Courtesy: Internet , Pixabay.com

22.1 Introduction :

Case 1 : You are visiting an ice -cream parlor in hot dry summer noon and you order a butterscotch with chocolate pudding ! You start enjoying ice-cream and after few seconds , you observe , the ice cream is excessively salty ! You cannot enjoy it after this observation and your mood to relax by eating ice cream in turn become excessively angry over the taste of the ice cream ! So , what was missing in taste of ice cream ?

In some ice creams ,to remain it cool for longer time freezing mixture is used in which salt is added to avoid quick melting of surrounding ice ! Somehow , this mixture enters into ice cream and as it is liquidous , it cannot be detected by open eye ! But when it is tasted , then people come to know about its real taste and get angry because it does not match to their expectation ! People expect that if it is butterscotch with chocolate pudding on it , then it must taste sweet -soft and coffee type taste ! If they do not get this taste ,the product does not fit to intended purpose and hence it is rejected by customer ! So ,morale of the example , ice cream maker has to test the ice cream for its taste , color

, pudding , degree of solidification of ice cream and its smell ! If these things are checked before serving the ice cream , it gets sold without any worry ! This is what importance of testing ! Testing ensures quality delivery of product & services !

Case 2 :

You are visiting a manufacturing company for final inspection of your product and while passing through that area , you find number of tests are carried out in that shop for different jobs as per their stages !

You observe some people are checking dimensions , some people are checking process defects , some people are checking machine parameters , some people are engaged in non-destructive test such as radiography , ultrasonic, dye penetrant , eddy current , some people are carrying out destructive testing , some people are carrying of weight upliftment trails , some people are packing material for safe transit !

What is your first impression after seeing these active visuals in the shop ? The first thought which come to your mind is my job is also passed through such types of tests and now I would like to see how my product meets its performance before dispatch from shop ! It will help me to install easily in my shop and run safely . Secondly , if I want to cross check any test value , I can immediately go to job and carry out random inspection of any dimension to find out the actual written value. If this value match for such type of 10-20 observations , I can get assurance that testing is done before showing this job to me ! This builds faith with customer and business relation become strong ! This is what the importance of testing and regular testing in any industrial environment !

Case 3 :

A product is delivered to site and its assembly is about to complete . Before taking its field trail for actual process , the product assembly need to check for leak proof ness ! Leak proof test of individual parts is done and it is certified but leak proof testing of assembly is not done and it is to be done after complete assembly at customer end !

During assembly when part manufactured at one end is assembled with part manufactured at other end , the connection is observed with minor gap ! A matching gasket is fitted within two joining parts ,but still the leakage is not stopping ! The leak proof test cannot be taken because pressure is dropping because of this leakage ! In such situation , how to test the whole assembly without attending this leakage ?

Team gathers together and a fitter present on site is advised to open that respective connection and check the straightness and water level of mating parts . If there are excessive shallow and bumpy regions , then with smooth filing , you need to level the surface properly so that mating surfaces remain uniform ! Same thing to be observed for other mating part and its mating surface to be made in water level . The third thing is you have to check the thickness of gasket and its pressure retaining class ! If you find supplied gasket has less thickness and strength than its intended design requirement , then how this type of gasket will provide total air tightness and leak proofness ?

Once the experienced fitter observes two mating parts and intermediate gasket for its specification , he observes there is change of class of gasket and hence this issue is observed ! So ,whether small gasket can cause assembly leak test at halt ? Yes , this is what mechanical engineering is ! You have to fit parts within given tolerances . If the dimensional tolerances are not maintained , then at site assembly such leakages produce delays and defects !

The experienced fitter level up mating surfaces with smooth hand filing and he changes gasket as per available gasket class ! The gasket got misplaced and hence low thickness -low strength gasket got fitted with mating parts ! This fine error is removed and part assembly is again tested ! This time , the joint was water tight and pressure required for leak testing was also raising as per standard observation ! The assembly is hold under pressure for thirty minute and no leakage and no pressure drop is observed ! At the end of thirty-minute , pressure is slowly removed and again any leakages are checked . There was no leakage and hence test proved to be successful !

Once this leak testing is done , next day field trial with actual fluid feeling is taken and the assembly performed swiftly to intended performance requirement ! This is what the importance of testing in any product manufacturing ! Let's look into various types of tests applicable to different products .

22.2 Raw Material Testing :

Raw material is tested before its production uses by physical inspection and chemical analysis . Destructive and non-destructive testing applicable as per design specifications is also checked ! Raw material is supplied with manufacturer's test certificate ! Some of the values are re-confirmed by carrying out production coupon testing in which parent material is joined by welding . Here strength of weld joint and strength of raw material is verified with respect to observed tensile strength !

22.3 Non -destructive testing :

Non destructive testing is carried out to ensure soundness of parent material and joint with the help of some volumetric test in which parent materials internal structure remain unaffected . Typical non-destructive test includes dye penetrant test , magnetic particle test , ultrasonic test , radiographic test, eddy current test ! The main purpose of these tests is to find out the discontinuities in the material and also to find out defect in the material !

22.4 Destructive testing :

In destructive testing discontinuities and defects in the material are tested by permanently braking or altering the test specimens ! The purpose of destructive testing is to find the maximum strength to withstand various stresses which can be static or dynamic ! Typical destructive test includes – Tensile test , impact test , fatigue test , stress rupture test , bend test !

22.5 Pneumatic Test :

Pneumatic test is carried out before carrying out hydraulic leak test ! In pneumatic test, containers or products are filled with compressed air and their inlet and outlet pressure is measured by attaching pressure gauge . The pneumatic pressure can be 2-3 Kgf/Sqmm and it is hold for that pressure for 5-10 minutes ! If there is no pressure drop for this time , the test is assumed to be leak proof ! A soap bubble solution is also applied to check whether there is any air leakage . Air leakage form soap bubbles ! More the leakage of air , bigger bubbles are observed . Noting the location of these bubbles , when test is over , the parts are repaired and retested till pneumatic test become leak proof ! If pneumatic test is not carried out before hydraulic test , then if such type of defects are noted during hydraulic testing ,then you have to empty the whole testing part and again move to repair area to carry out the repair after complete drying of part ! This takes more time and hence to avoid this risk , pneumatic test is done before the hydraulic test !

22.6 Hydraulic Test :

As discussed earlier , hydraulic test is carried out by filing plain water inside the testing vessel or container ! After complete closing of all openings and tightening them properly, the vessel is pressurized to applicable hydraulic test pressure slowly and firmly ! During pressurization, there should not be any pressure drop ! The pressure must raise swiftly and once it reaches required test pressure ,it must remain stable for test observation period !

In the observation time of 30 minute, if there is any leakage, water comes out from such joints ,these joints to be marked and repaired after test by moving empty & dry vessel to repair area ! After repair, rehydro to be taken to ensure its leak proof ness !

22.7 Microstructure testing :

This is a typical lab-based testing and here material microstructure is checked with respect to its standard microstructure to observe number of phases present in that material ! Here sample removed from identified section is cut to suitable sampling size and polished to provide extremely smooth surface . It is etched with suitable etchant and at right magnification , its structure is noted ! ✱✱✱

THRILL 23 : THRILLING DESIGN REVIEW

Image Courtesy: Woman , Pixabay.com

23.1 Introduction :

The very first pictorial representation of any engineering part is possible only because of the scientific art of making imaginary sketches on paper using different aspects of two dimensional and three-dimensional view !

Engineering Drawing and Design is imaginary yet mathematical concept ! One can easily draw a tree with his or her imagination but when someone wish to draw a house for permanent living, then he has to imagine how he will protect the house from hot summer , rainy season , chilling winter , earthquakes , floods , thefts and encroachments ! Accordingly, they have to design a master plan in which practical layout is drawn to show the position of rooms , position of entry and exit , open spaces and security gate ! This type of drawing based on considering actual site condition is known as engineering drawing !

Standard work involves reputation of same design for 'n' number of times ! So , if you have to construct million houses of same design , you simply prepare a single drawing of allowable area and get its approval after proper scrutiny and then you simply go on repeating same task till you reach millionth house plot ! This is what scalability of the drawing !

So , designers have different ideas as per varied customer requirement , they can show single drawing to million customers or they can design million designs for million customers ! After all its pure work of imagination supported by mathematics !

So, in this discussion , we are spreading light on thrilling design review and which point to be considered when you are carrying out drawing review .

23.2 Design Calculations :

Design calculations is basic check point of any engineering product as they support strength and durability of that product ! If a product is expected to serve for typical service period , then it must have required strength , applicable factor of design safety and relative flexibility so that customer can use that product according to his work load need !

So, design calculation needs to have detailed mention of maximum performance limit of that product ! Next to this, design calculation must mention maximum allowable performance limit of that product ! After this mention, design calculation must mention normal working limit under which most of the time the product will serve ! When these limits are specified, one can understand how careful the design is made to ensure total safety of product and its surrounding !

After safety and capacity determination, design calculations provide dimensions of each part according to three-dimensional requirement . These dimensioning ultimately finalize the size and shape of the product !

Design calculation is supported with proven formulae of typical engineering product ! When you are designing a rotating fan ,you need to take into consideration the electrical requirement of assembly so that variable r.p.m. can be produced with change in current . At the same time, you have to design an electric shock proof fan ! It must have safety cover on front and rear side and it must have a connecting wire from where required supply can be given !

When you are designing a tank , you must know the capacity of tank , its connection requirement and its opening , size of every opening , height and weight of the tank , you need to know the thickness of the tank which can withstand pressure and wall stress caused by fluid inside the tank ! Accordingly, you will use formula of volume calculation , minimum thickness calculation , stress distribution across walls , maximum fluid carrying capacity and maximum and minimum fluid level ! Then you will determine network producing volume and safety volume of tank ! What will happen if tank overflow because of excess amount of fluid , how the alarm is generated when the tank is crossing safety limit ! These points will be noted in typical design calculation of tank !

Design calculation will tell you whether the designed part is able to carry applied load , it will tell the design is safe , it will record maximum required thickness and provided job thickness for every part , in all such cases ,provided thickness need to be higher than required thickness to conclude the design is safe for use ! Once you check performance requirement related calculation accuracy , your next job is to ensure correct dimensions are written for every part to get that performance during actual operation !

For example, what will be the mass of solid material filled inside a tank which is 10 x 20x 5 cubic meter and the density of material is 1.5 Kg/ Cubic meter ?

Here , we know ,

Density = Mass / Volume

1.5 Kgf/Cubic Meter = Mass (Kgf) / 10x20x5 Cubic meter

Hence Mass (Kgf) = 1.5 Kgf/Cubic Meter x 1000 Cubic meter

Hence Mass = 1500.0 Kgf

So , any mass filled up to 1500 Kgf will be occupied inside the tank . If you increase the mass even by 100 Kg , it will overflow the container ! If you fill just 150 Kg mass , then it will occupy 150/1500 x 100 = 10 & of its design capacity ! If you fill 300 Kg , it will occupy 20%

capacity , if you fill 600 Kg then it will occupy 40% capacity , if you fill 900 Kg , it will occupy 60% capacity , if you fill 1200 Kg , it will occupy 75% capacity and if you fill 1100 Kg , then it will occupy 91.66 % of its capacity ! So accordingly, the tank will be marked from outside and inside to show its all-possible capacity limit with which it can work safely !

Based on capacity limits control knobs are provided so that typical level indicator will raise alarm on reaching various capacity limits ! Here one has to just keep adding the material , once the expected performance level is reached , the alarm will give the required signal !

This is what the example of design calculation is ! Designer decide maximum capacity and accordingly they provide number of options on which customer can use that product . They also provide alarms with which customer can get required details ! Customer has to pay for this safe and secure arrangement by procuring that part ! This is the simplest example of design calculation ! When you are designing a static and dynamic part , the design calculation involves study of product under static loading as well as dynamic loading ! As we know , dynamic loading exerts severe pressure and stress on part and parts are liable to damage early ! To avoid this , dynamic loaded products to be designed with maximum applicable dynamic load and certain technical allowances which adds to strength of that product ! When a dynamic part works in harsh environment , then that corrosion allowance also needs to be added ! Thus, design calculations take into account typical service environment in which product will serve and everything later starts accordingly ! May it be selection of material , it's costing , its thickness and size , its way of manufacturing all decided by performance expected in given service environment !

A part designed for harsh environment with high safety provisions can withstand any regular service environment but a part designed for regular environment fails quickly in harsh service environment because of lack of provision of various allowances applicable to that harsh environment ! That's the reason , many products failure are noted when atmospheric changes happen suddenly ! The selected material cannot withstand environmental impact as that much safety is not considered during product design ! When designers are not clear for safety requirement , they can use guidelines of international codes and standard about typical product design ! These codes are formed after discussion , voting and approval or reputed technical bodies and field professionals as well as statutory officials who are responsible for general public safety ! This also ensure , the standardness of product design irrespective of country of its installation !

23.3 Dimensioning :

For every product ,minimum three views are required which are front view , top view , side view ! This view must indicate major dimensions of each part ! If there are sectional views then the other view must indicate the arrow head and direction of that section ! These points to be critically checked during drawing review !

23.4 Detailing :

When you cannot explain everything on single sheet , you attach second sheet of minute details . Detailing involves start to end mention of typical part making !

23.5 Bill of Material :

Bill of material mentions developed sizes of every part, its material grade , its quantity , and the material strength as per part number identification in title segment !

23.6 Notes , Title and Design Data :

Apart from main drawing and detailed drawing , design data , mandatory notes , drawing scale , title block mentioning product details and customer details , drawing number and revision number , footnotes and reference drafting file name , special instructions , applicable tolerance tables are specifically need to review, to find out customer requirement are included in drawing correctly ! There is authority approval table in which you have to sign once you review that drawing !   ✸✸✸

THRILL 24 : THRILLING ORDER REVIEW

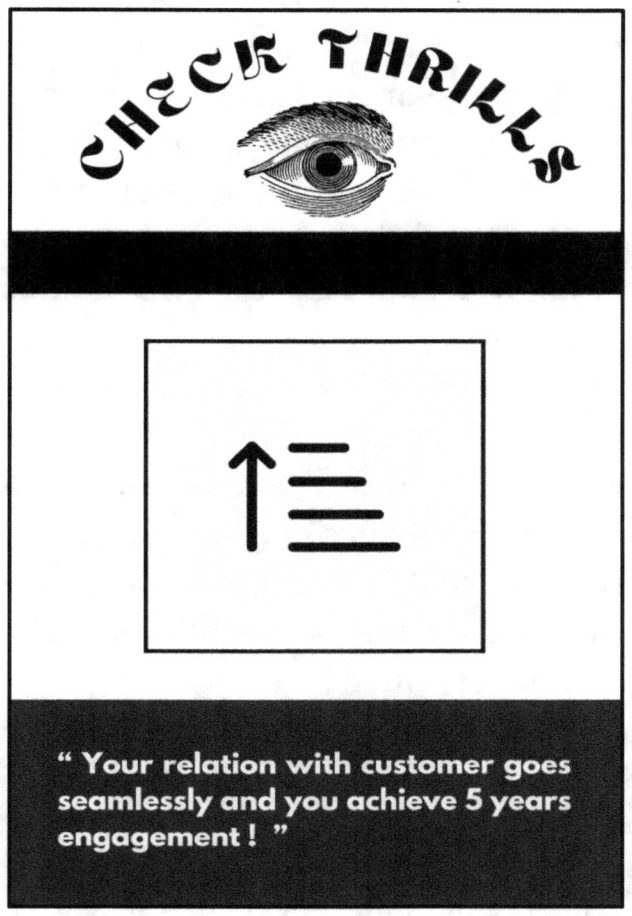

Image Courtesy: Sort , Pixabay.com

24.1 Introduction :

Case 1 :

You are visiting a food mall on highway during your regular travel to busy Mumbai – Pune express way ! You daily go to Mumbai , complete your work and return same day in the evening ! Your travel time is around 7 hours to and fro while your daily work takes 4 hours stay in Mumbai ! You drive with your own car and driver and you both take first break in this food mall and while returning also , you take the break in same food mall ! This is your routine for last 5 years and food mall manager and staff know you personally because of day-to-day interaction ! You have given them standing instruction about breakfast and the instruction is 'Give us different variety of breakfast every day ! Avoid repetition ! '

So, how the busy food mall management is responding to this standing instruction since last five years and maintaining sweetest relation with this regular and repeat customer ?

The thing is simple ! Staff has weekly menu plan in which every day they prepare minimum seven new items , so to meet this customer requirement , they have to avoid item which is given last day ! If they exclude this item, then any item from rest of six items is welcome to this customer !

They have a digital billing system in which programmable customized billing can be made ! For this customer , they have allocated a permanent customer number and in the menu list they have added a programme which exclude 'X-1' item and accept X , X +1, X+2,X+3,X+4,X+5 items where X stands for that day's billing ! It means , if it is Monday , the programme will exclude item served on (X-1) day ,that is (Monday-1 day) which is Sunday as feed in the programme logic sequence ! The rest of any item is acceptable to customer and they can serve any one out of available list ! So , with this programme , for last five years , they have same billing system and staff serve the food without any instruction ! Customer enjoy the delicious taste and pay the monthly charges in advance and moves to their journey within ten minutes !

So , why so much time consciousness for this customer ? Customer has made such type of deal with regularly visiting food mall to save time in traffic once they reach entry point of Mumbai in the morning and entry point of Pune in the evening ! If they could complete breakfast in ten minutes time , they can avoid traffic and reach comfortably to their work place and can complete that day's target within four hours by non-stop working ! As soon as their daily work get completed, they return to home again stopping at same food mall !

So, what are the takeaways from this simple daily life example ?

1) Your product meets customers daily necessity and for which he pays you monthly advance price !
2) Customer has very less time to be with you for that necessity and he has some different deliverables !
3) All he requires is, you must follow his one-time standard instruction and fulfill it when he is there !
4) There are no changes in customer requirement and you have to fulfill daily requirement within least possible time !

5) Your relation with customer goes seamlessly and you achieve 5 years engagement !
6) To avoid any mistake in standard instruction , your staff prepare a customized billing programme which takes care of ' different menu every day and avoiding repetition of yesterday's menu , rest is acceptable ' !
7) The technical team link this programme to your menu list and confirm this programme for one year's calendar for which the software is designed . Next year , you will amend the programme or you can continue the same program with updating the setting !
8) Now irrespective of staff , customer receives his order on time every day , two times a day and he is just happy with your service !
9) Looking at his way of customized service , many customers approach the food mall and, in the end, food mall allocate a special dining arrangement for customized orders which are pre-booked and just waiting for arrival of their permanent customers ! On second side ,the other new customers are welcomes and served as per their different orders which are not programmable !

So, this is what typical product and project environment is and this is how the thrilling customer order requirements are dealt by management so that seamless business transactions keep happening ! In this discussion , we are going to spread light on different aspects of order review !

24.2 Basic Customer requirements :

You have to find out what are customers basic requirement and is your current product model or range of available models meet those basic requirements ! Once you review customers basic requirement , you suggest most suitable product model to customer . Customer goes through it and if he like that product , he purchases it !

In case , if he doesn't like that product , you have to show them other products , if he finds the required one , he purchases it else he leaves your site or shop !

Here , you need to be aware of the fact that if the product is not available in your shop or your site , then it is not available anywhere in the city ! And to have this knowledge , you must know your other competitors in detail ! You must visit those places and see which products are served by them and at what price ! Are they providing any type of discount so that customer is going there ? You also have to visit your city's biggest and smallest distributors to check whether any new product is arrived in market and if it is then you have to purchase sample stock to see how customer look at it ! This small point comes into reviewing customers basic requirement ! If the product is not available with you , it means , it is not available in any other shop or site in the city ,because you are the sole manufacturer and supplier of that product !

24.3 Special Customer Requirements :

Customer tell you special requirements for which he is ready to pay extra price ! These special requirements are nothing but additional safety and performance features expected in your product by suitable customization !

If your regular thickness is 10 mm , customer will ask for 14 mm and he is ready to pay the differential price ! If the original size is 100mm , he may insist for 200 mm to increase the production at his end ! If this dimension rise is technically feasible , then you can accept and fulfill the order ! Else you have to tell the technical restrictions on such changes !

This customization gives you excellence in your product design ! There are two things in dealing with customer ! Either you have to provide standard job to customer and let him use as per his need ! In other case , you have to make your job as per customer mentioned changes and gain the additional profit margin for degree of feasible technical customization ! Technical feasibility of every customization is very very necessary ! If technical and safety related parameters are neglected because of repeat follow up from customer for unfavorable customization , then there is going to be huge risk to customer and your own business on supplying technically non-conforming part !

So , every order reviewer has to see what customer is exactly asking which is not present in standard work order and how they can provide that special requirement without affecting its technical requirements ! If they can provide special requirement without affecting product safety , then the order is acceptable , if it's not , then you have to communicate in written to customer about how much customization is possible safely ! This is how special customer requirements are dealt with !

24.4 Price Variations :

Prices keep changing and you have to convey the price change accordingly to your repeat customer ! A fair estimate is given by you to your customer after enquiry from their end and in this estimate, customer try to make few changes from his side ! You have to work out the calculations with which these changes can be accepted ! If your present profit margin is going down to intolerable level , then you have to tell rock bottom price from your side ! This new offer will be reviewed by customer and he can give one additional order with asking some type of discount ! You have to again review the new offer with additional product purchase and you have to see in one production run you can complete both orders in X amount of extra cost , so making these two jobs one time is profitable and even after providing the discount , the deal become favorable ! Thus ,you accept that order with price variations! This is what key points of order review ! Order reviewer has to make all attempts to meet customer demands in techno commercially feasible way !

24.5 Tighter delivery requirements :

Some orders are given by customer just because of noting your overall preparedness ! Customer knows you can fulfill this order in record time without any rework or delay and hence he keep giving more of such orders !

As an order reviewer , you have to talk with your customer and ask for annual requirement of such fast-paced jobs ! After noting this requirement , you have to develop your production capacity and keep sending those products to customer !

Even if a new job comes with urgency , just by ensuring with your production people , you can take that order firmly !   ✸✸✸

THRILL 25 : THRILLING ON SITE CONCERNS

Image Courtesy: Cranes , Pixabay.com

25.1 Introduction :

Case 1 : You have bought running water tap from nearby hardware store and you replaced with your old tap which is causing water leakage ! For two to four days , that tap done its job and on fifth day ,it started showing leakage ! You visited hardware store again and communicated this concern to shop owner ! He replied gently , Sir , it may happen for 3-4 in 1000 ! You give us that tap and take a new one ! For your additional knowledge , we will provide the most recent tap which comes with QA approved certificate which mentions water leak test results specifically as a new additional feature of this product ! This issue is now resolved at source and new products are coming with this new feature !

You see that product ; its certification and you get that tap ! You fit it gently and observed for one month . The tap is running smoothly and there is no water leakage type of thing . Secondly , you also observe that there are some changes in shape and size of this tap and it can be rotated very freely compared to earlier operations ! You are now happy with the service provided by water tap and you are pretty sure ,this tap will provide its service for long time ,which can be 5-7 years !

Case 2:

You are a manufacturer of welding wires . Your shop makes monthly production in the range of 50 tons ! These welding wires are packed in welding spool and supplied to local manufacturers for their welding need !

There are near 100 customers and everyone purchase your product in the range of half ton per month ! Twice in a month , you visit each customer personally and take their feedback about experiences of their product ! You also ensure the mechanical properties of as welded parts are satisfactory and in no test weld metal is weaker than parent metal ! You also verify the percentage elongation obtained in welded part specimen is within permissible limit . You also observe face bend and root bend results . You check impact value of as welded specimen ! You verify microstructure and heat affected zone of welded sample and thus you collect data from all 100 customers !

You share this data with respect to supplied materials batch number and order number and co-relate with your lab testing values ! To your surprise , you observe , almost 99 % lab results are matching to those observed at 100 customers ! The remaining 1% results are also not so different . They are observed within acceptable tolerance limit of specified test values by respective welding standard ! But the one thing that satisfies you is, the field test results carried by 100 customers and lab test result carried out by lab staff are same to 99% !

This is what the role of testing laboratories in creating products that suit to intended purpose ! Lab certifies product after due verification of each and every standard requirement . If any deviations are noted , then they are compared to applicable tolerance range , if it doesn't fall in applicable tolerance range , they are either rejected or they can be used for low strength standard which is available !

Testing laboratories are staff working for customer and employing organization whose basic role is to ensure safe release of all products for any customer !

So , what one can learn from these two different cases ? In first case , customer complains about the product and then he gets the replacement . In second case , the manufacturer himself visit 100 customers to see his lab results are matching to 100 customers filed applications of his product ! The observations establishes that results are conforming and there is no issue with material supply !

So , its manufacturers choice and interest that whether he would like to perform pro-active fact checks of his creations or he want to perform reactive post-analysis of his sold-out products ! A growing business always work in proactive way and reduces amount of site complaints !

In this discussion , we are spreading lights on thrilling on site concerns and how the product quality can add value to reduce the majority of these concerns !

An ideal manufacturer is one who provide the top-quality material in quickest possible time at lowest price ! An ideal customer is one who pay the highest price for top notch quality and allows required cycle time to manufacturer !

The whole business reciprocates within these two perceptions of ideal customer and ideal manufacturer ! One who matches perfectly become customer friendly manufacturer and grow substantially ! Let us see , number of on-site concerns .

25.2 Material Concerns :

Material supplied on site found substandard ! It doesn't meet required strength concerns and breaks with minor loading ! Why this type of situation occurs ? The basic reason is not customer nor manufacturer is aware about the standard material required for that particular service requirement and they use low strength material for high strength requirement ,which get failed when performance requirement crosses ultimate strength of the product ! Hence , that material is not wrong , its selection for intended application is wrong ! Material has served well within its allowable strength limit but when strength requirement raised , it couldn't meet that requirement ! Hence , before finalizing any order , all service requirements are taken into consideration and then only suitable material to be used for manufacturing that product !

25.3-Dimensional & workmanship Concern :

All product drawings are made with the help of latest design software's which are capable of designing part on large scale as well as design for minute details ! During expansion of main drawing to detailed drawing , all dimensions are transferred from main drawing to detailed drawing . In this transfer , designer has to take care that there should not be any deviation in between main drawing and detailed drawing . If there is difference in both dimensions , then during processing people will get confuse and ask for clarification which will consume productive time and people from all production line will ask this question whenever they face same problem again !

Also , a typical assembly can consist of many interconnecting parts and even one part gets that dimensional error, then it causes whole assembly to mismatch on site ! Many times, parts are manufactured at respective work places , checked with respect to individual drawing and sent to factory dispatch to see the clearance report before sending to customer ! In such case , sometime assemblies are done in shop to see the correct matching of all interconnecting parts received from number of suppliers . If there is any issue , same is shown to designer and respective manufacturers of that faulty part and necessary correction is initiated so that this issue will not repeat in next series ! In case , such type of defect identification is missed , then same get highlighted during assembly at customer end and then concern manufacturer has to provide justification of this error to all stakeholders !

Such type of dimensional concerns can be avoided during drawing review . You have to check whole assembly drawing and each individual part drawing and see the matching of required interconnecting dimensions. If there is error , you will immediately note it and then you have to give necessary feedback and get that work done !

Dimensional concerns are attended by changing current dimension to required dimension !

25.4 Performance concern :

Every assembly has typical performance pattern . Assembly is expected to work seamlessly at low load , medium load and high load ! When the load variations are done , the performance is expected to increase than earlier one . There is mathematical equation to every performance variation ! Performance must change according to its governing technical equation around which the construction parameters are set ! If assembly behaved erratically and illogically , that assembly is considered as hazardous for use and it is restricted for further use !

So ,typical performance issues observed are related to smooth starting , swift closing , intermediate status changing from one state to other , control related , speed related , cleaning related , jamming or interruption related ! The main reason of such concerns is non equilibrium of available variables and constants ! When the variable and constants are in perfect state of equilibrium , any performance variation obey that equation and hence every change is controllable . So , when such type of concerns is recorded, engineers have to find out static and dynamic performance and find out the missing link or missing variable which is not in equilibrium state and the reason behind its disturbance ! Once you resolve this error, systems functions in equilibrium and the variation become predictable and certain !

25.5 Safety Concerns :

Safety concerns are observed when parts have some type of hidden defects which get open because of typical service environment ! Typical safety concerns refer to fluid leakages, sudden expansion or contraction of working parts, failure of parts because of fluctuating loading, man handling of delicate assembly and its control, lack of proper knowledge of risk mitigation for newly installed assembly ! Such concerns are corrected by respective site visit and site action plan execution !

25.6 Financial Concerns :

They deal with per day or per second losses ! ⊛

THRILL 26 : THRILLING NEW PROJECT

Image Courtesy: Pennant , Pixabay.com

26.1 Introduction :

What is the most exciting part about any new project ?

People are excited because of a brand new concept understanding , they are interested to work on this new project which is going to change the way people earlier used to work -feel -earn and enjoy , they are excited about new additional features they are receiving in same price , they are excited for extremely chick looks and size of the product that simply pulls customer towards it , they are eager to know how many new technologies will be incorporated in this project environment and because of which how simple the job will become ! New projects simply bring new hopes on which people love to rely , love to work with utmost dedication and love to tell everyone out there in the team to just focus on results ,because these results will set record in this technological revolution and you will be the pioneer of this innovative model as whole !

In this discussion , we are going to spread light on thrilling new project and how they are tackled as such !

26.2 Generation of new concept :

They say , ' Need is the mother of new invention ' and so is the thing about generation of new project concept !

Business & market are extremely volatile and risky subjects . If you have to remain in the business for long term along with your internal capabilities ,you have to ultra smoothly blend with your external environment ! Then only the business gets covered with soft blanket of internal knowledge with wisdom and external order flooding with profit marathons & growth hurricanes ! Really !

If you look at the current new products just like electric car or digital payment mobile applications , you will observe , in fuel market ,fuel prices are continuously rising and international conflicts are just making scene difficult . Secondly , the natural fuel sources are also depleting and on one fine day a stage will come where there will be no oil present in this world ! If this happen then how the people will move to places ? What can be alternative for depleting fuel resources ? When this type of thinking is started , researchers started thinking about new concept of electrically charged car which can run for long distance ! When technical analysis of this concept is started , people started brain storming and experimenting ! Now a stage is achieved in this direction where the developed electric car is available for sale ! The work of installation of charging facilities is also on the way and down the line of next three to four years , a greater number of pollutions fewer electric cars will be seen on the road and thus issue of fossil fuel scarcity will be handled with technological option of electric car !

Electricity is fairly available in nature and it can be produced by water , wind , thermal energy , solar energy ! So , huge requirement of electric charging car can be addressed sufficiently by various options of electric energy ! Right now, Lithium battery is in focus for charging the electric car ! So , this is how the new concept get generated out of market need and future risk !

Secondly , when information technology is evolved , many students learnt those ways with which it can become useful to humanity with the help of excellent coding and programming !

Engineers noted that general banking carry out two basic mathematical transactions , which are debit the money and credit the money ! So, there is alphanumeric system which takes care of this transaction . The person has a bank account and he transact the money through this account to other bank account ! Then what will happen if human interface in banking network is assisted by digital technology which will store every account detail and its transaction details with authorized servers . This development given rise to facility of e-banking !

With e-banking banks are connected to each other and transactions started with the help of technology ! Here the limitation was use of laptop or desktop was necessary and people need to carry out transactions with the help of computers only , no other gadget was useful in e-banking development phase !

With evolution of smartphones supported by suitable operating systems , these smartphones are converted into microcomputer up to allowable extended memory and RAM ! This technological marvel helped engineers to develop mobile applications that can be used by people on huge scale !

This given birth to connecting two banks with respect to registered mobile number of their customers ! So , here transaction is happening from one customer registered mobile number to his authorized bank linked to that number and banks internal system used to connect with recipients' banks internal system and recipients bank used to receive that money from sender bank after sender request and used to add in recipients account and finally recipient bank used to provide that money to recipient in digital form ! Means the physical currency got replaced with digital database where every account is balanced as per respecting authorized banking transaction and authorized safe banking limit !

Digital banking reached to retail business with the help of portable and handy QR code scanner and this further simplified payment starting from Rs 1 to 50,000 instantly ! This invention changed the way people used to transfer money and it was protected by mathematical foundation and hence proved extremely safe ! In any adverse event , money used to re deposit in account after few allocated days by the banking system !

So, here the technological benefit is used to have new concept of financial transactions using mobile phone ! Smartphone !

26.3 Modification to existing design :

This is secondary development in new product design ! In this type of new products , you carry out certain number of new design changes to improve the existing performance of your old product , you may add some more facilitation to record the things accurately and hence control the performance more smoothly ! Such changes keep happening regularly and your basic product design remains almost same . Whatever changes are done in this type of new project are related to additional features !

26.4 Modifications after constant site issues :

These types of new projects are developed after receiving customer feedbacks from various sites ! You have developed a product in controlled specifications and dimensions but site conditions are changing haphazardly and your product is losing its strength ! So, customer give you feedback about that particular environment and ask you to develop a solution that will face that environment swiftly ! This feedback give birth to development of new project suitable to customer specified environment ! The product is identified by that environment easily recognizable feature !

Suppose a customer ask you that your products height is causing some visibility concerns and hence to suite the same design volume , can you reduce height and increase the length and breadth so that we can easily view the product at our existing setting else we have to carry out expensive arrangement to set your product in this facility and hence we may think to look different option ! When you listen to this customer requirement , you analyze other similar customers issues and thus noting the frequency of this feedback , you change dimensions accordingly and launch a new developed product ! You rename the product and tell your customer about it !

Some customers need extra expandable capacity ! It is like typical railway engine like design ! Customer expect that you should design such a strong capacity engine that we can attach additional bogies as per our production pressure ! In short, product design needs to be sturdy and robust which can handle expandable additional loading as per production requirement !

Some customer gives feedback to change aesthetics of the existing product ! These changes may change the overall product appeal and appearance which help to attract more customers ! But these types of changes are just supplementary ! Customers expect performance improvement developments regularly to cut down their costs and raise bottom-line by using your products !

26.5 Modifications after statutory & international changes :

Do you know, on one fine day, some countries directly ban typical material exported to their country ? Do you know, on one fine day, few countries charge hefty penalty for not adhering to their latest environmental concerns and amended product protocol ? Do you know , some countries sue the manufacturer in international judicial forum for not conforming to updated pollution emission norms ?

Friends , if you are doing regular business with other nations , you have to establish a sales and marketing regional office in that country and make sure the technical and legal requirement of that country are met in your new order discussion meeting ! Once you get detailed written information about statutory requirement , you have to get that job inspected from authorized inspection agency and show its acceptance certificate to overseas customer ! When this technical faith transfer happens smoothly , dealing with export orders become easy and easier ! Same thing is applicable for fines , penalties and judicial complaints ! You have to specifically note the conditions of default and applicable penalty ! You should ask for freehold extension in the case of adverse events before finalizing order ! The duties and penalties should be proportionate to profit from order and jurisdiction should be easily accessible ! ✷✷✷

THRILL 27 : THRILLING MACHINES

Image Courtesy: Gear , Pixabay.com

27.1 Introduction :

The job of inspection is done in three different work set ups ! These work set ups are open material yards , protective workshops and customer sites where your product is installed ! Surprisingly , at every location , there is presence of some thrilling machines which are busy in carrying out their job at dedicated workstations and hence you need to be totally aware about the personal safety measures before approaching nearby these machines ,so that you can learn more about them in static and dynamic condition and thus carry out your inspection safely by following safe working practices !

In process inspections are typically required to take this extra care . The machines are in full swing and you are witnessing the outcome of process standing in nearby safe region ! In this inspection ,you view total process path starting from input entry , processing mechanism and exit rhythm ! If this is happening as per process synchronization , you get accurate process result . If there is any process deviation , the machine performance starts showing inferior product quality and thus in 'in process inspection' , you have to search such process loopholes and plug them before further inferiority is created in final product ! For this sole reason , in process inspection is done !

In this discussion , we are clubbing the topic of thrilling machines along with key aspects of in process inspection , which is considered as most proactive stand an inspector can take to curb rising process rework in any workplace !

They say , if you have to check how hot a surface is or how sweet the sugar is you have to touch the hot surface with your hand or check with the help of thermometer ! In case of sugar, you may use glucometer or your tongue ! Then only you get the real idea of overall process result ! In the same way , if you want to arrest process deviations you have to either install a full proof mistake proofing mechanism such as 'POKA- YOKE ' or you have to stand in front of running machines for number of hours till you record all abnormal indicators observed while producing output for plant !

The typical preventive and corrective maintenance do carry out such type of machine inspections regularly, still as an inspection engineer, many time you also notice typical malfunctioning of machines which are responsible for production of poor quality ! And hence this machine operating knowledge is basic to investigate root cause analysis of major quality deviations ! Because, in the end, every root cause analysis has single or multiple reason out of 5M of managements which are Material, Method, Measurement, Manpower & Machines ! If you can add sixth M which is Money, this can be hidden reason of process defect ! If sufficient money in not invested in establishing the quality management system in the start, you are going to take risk of inferior production after certain time of three to five years. This is because, the low-cost machines will become risky in functioning after certain time because of low strength of its building elements ! So, let's discuss various types of thrilling machines and how in process inspection is carried out by standing in front of it !

27.2 Plate Cutting Machine :

On plate cutting machines , plates are cut to required size ! Plate is procured in standard length of say 2.5 meter by 10 meter and then after calculating the required length and width of the part , the plate marking is done by plate cutting machine . In the vacant plate , if there are parts of same thickness are available , then their placement is done in available area and this whole fully marked plate is cut by plate cutting machine . Every part is punched before cutting with permanent stamping so that once parts are removed , they can be easily identified and supplied with respective job trolley or job cart !

During plate cutting in process inspection , you have to note the plate thickness and plate size is correct . Its material test certificate supplied by mill is available and the profile cutting layout is marked properly . In inproces inspection of marked plates , you measure the part dimension before giving cutting instruction , this is because , the raw material is highly expensive and if it got cut less than the required size , it become useless for that part ! Neverthforth , its sizing become difficult if it cannot be used for other jobs ! Hence precutting in process dimensions check of plates ensures accurate cutting and accurate assembly later !

27.2 Plate Rolling Machine :

In any manufacturing organization , this is an important machine which is used to prepare round cylindrical surfaces by rolling the plates in between two revolving roles . The diameter and ovality of roles are checked with respect to profile templates . Every rolling machine has a control panel from where the speed of roller and load taken by rollers is synchronized so that machine keep working without any stoppage ! When the shapes are welded , they are rerolled to remove the distortion observed because of welding !

During rolling in process inspection , you have to see part number and job number and its stamping on plate on visible side (Both outside and inside) ! You have to note thickness of plate before taking for rolling . You have to check length and width of plate and it is according to dimensions given in bill of material table ! After rolling you have to check out of roundness of cylinder and plumb so that X & Y axis diameters are within plumb ! If you notice any not acceptable ovality , you have to again reroll the cylindrical shells till you get accurate roundness ! Shells form basic enclosure of further mechanical equipment's and hence its surface inspection both from internal and external is must to ensure defect free plain surface !

27.3 Longitudinal and Circumferential Welding Machines :

The work of joining shells is done by welding machines . Shells are joined by longitudinal joints and circumferential joints . The welding can be done by portable welding machine such as SMAW and FCAW technique or it can be done by semi-automatic SAW technique ! In SAW technique , longitudinal welding is done by placing joint in gravitational 1G position and the welding arm carry out welding from one direction to other ! In circumferential joint , the shell is rotated on rotator with extra slow speed and the machine is kept over the joint in 1G position by using suitable safe fixture . Here welder is sitting on job directly to weld in 1G position for better weld penetration and hence job movement is to be done only after his instruction to avoid his falling ! Extra precaution is necessary in circumferential welding of higher OD shells ! This welding can be done in vertical position with welder standing in front of job and job is rotated as per progression of welding !

In this 'in process inspection', you have to see approved welder is welding the job, approved WPS is available nearby, approved PQR is available near machine, welding machine calibration certificate is available, welder ID is available, welding consumable certificate is available, proper blowers are placed at safe distance to protect from welding heat, safe access arrangement like ladder and platform is available for keeping machine on platform while welding on topside of the job !

27.4 Machining Section :

In any machine shop number of lathe machines, drilling machines, facing machines, pull bore machines, grinding machine are fixed to perform particular machining operation.

The process involves setting the machine to required parameters , set the part to be machined inside the fixed and movable jaw of the machine set the required machining tool with the machine tool setting holder , tight it properly , keep necessary lubricating coolants ready and start the machining operation !

The detailed machining drawing will be kept near you and you have to refer for exact machining dimensions , you have to first take a rough cut and see the exact available surface of part to be machined , once you get the idea of tool movement , you have to take finish cuts ! Depending upon the machinability of part and hardness of tool , the machining time varies ! You have to refer the best combination of tool for that material and keep working with it ! Always note , a hard machining surface breaks down soft tool easily and extremely hard tool used for soft material may make machining faster but it may harm surface adversely in the event of tool creating excessive machining marks ! Hence , tool selection as per machining part need to be taken care of ! In 'in process inspection' , after every cut dimension is verified by guiding pin ! Guiding pin is used as reference for cutting and it also alerts workmen to stop before cutting goes beyond guiding pin mark !

27.6 Hot Furnaces :

Hot furnaces are facilities which are made up of number of machines ! There is need of electric connection , there is need of mechanical arrangement to load the charge in furnace , there is need of masonry work , there is need of thermal systems , there is need of coolant supply ! So, here you need to know about many interconnected systems !

During in process inspection near hot furnaces , you have to see , for which purpose the furnace is exactly used for ! Whether it is used for metal making , heat treatment , preheating or for some other thermal needs ! Based on that need , you have to see standard operating instruction and the standard work done in that area ! You have to see charging , melting and pouring operations follow the composition and temperature limits and furnaces are properly calibrated ! Thrilling machines are backbone of advanced industrial progress ! ✹✹✹

THRILL 28 : THRILLING PUBLIC COMMENTS

Image Courtesy: Comments , Pixabay.com

28.1 Introduction :

Que. 1) How do you feel after eating in this restaurant ?

Customer 1 : Superb , super taste !

Customer 2 : Oh, I can't forget the freshness of food here and next time I will come with my whole gang of college buddies !

Customer 3 : Super excited to visit this place , this place come to know through one old friend , I was on official tour for some work , now relaxing with dinner here with my new friends !

Overall General Impact : Positive

Que 2 : What will tell to your friend about this new car ?

Customer 1 : Go and get it !

Customer 2 : Its new marvel and chase for it !

Customer 3 : New entry in car market , worth spending high price !

Overall General Impact : Positive

Que 3 : Why you will not recommend this product to your friends and dear ones ?

Customer 1 : Its useless ! Simple !

Customer 2 : Not sure , but it can breakdown anytime ,anywhere and repairs requires hell lot of time !

Customer 3 : Safety risk ! That's it !

Overall Impact : Negative

Que 4 : Which feature of this service never satisfies you ?

Customer 1 : Incomplete filling of required material at not so accessible parts .

Customer 2 : Loose packing of material in many joining recesses which result into loose joints .

Customer 3 : Surface finishing after material fill up is not look uniform and hence aesthetic quality looks poor !

Overall Impact : Negative

Que 5 : Will you buy this product again ?

Customer 1 : Yes , pretty sure !

Customer 2 : This product not fulfilled my total requirements . After using this product , I found other options available in the market and hence I am going to buy another product not this one next time !

Customer 3 : Not on priority , but when nothing is available , then I will keep this option open !

Overall Impact : Positive – Negative – Mixed !

Friends , welcome to the last few topics of this book ! We have already discussed the relevant topics which are necessary to understand thrill of inspection ! In this topic , we are spreading light on thrilling public comments and accessing it overall impact on the minds of current and future customers !

Public comments refer to various types of public reactions and responses received after using your product and services ! These comments are read by other users of your products and services and they may create positive , negative and mixed feedback on your further business growth ! So , let's see , step by step ,how thrilling public comment influence your business strategies and business decisions .

28.2 Loyal Customer :

These customers will always love to give positive response for your product and services . There is something in your product or service which is attracting them towards buying it again and again ! Some customers use typical product for number of years . Even there is modification in product and both old and new product is available , they stick to old product only ! This is what brand loyalty is ! These valued customers are your routine business generators !

28.3 Angry Customer :

Some customer ,which can be 6 out of 100 may not be totally satisfied with your products and services and they will just boast with furious reaction if you ask about their customer feedback ! You surprise to know that 94 customers are not showing so much dissatisfaction than these 6 annoying customers , what can be reason behind this agony and anguish ? When you give a deep thought to these complaints after giving due time for its fair analysis , you find that customer is not wrong in his feedback but the working conditions present in that area are not suitable for this product optimum uses and hence they need to recommend a new version of this product ! When you discuss the matter with your angry customer and show him free trail of 15 days in his shop , he surprisingly accepts the new product and stick to your organization for rest of the business affairs ! You need to know the knack of fulfilling requirement of angry customer by providing him enough product options suitable to their present work conditions ! This is what business diversification and flexibility is !

28.4 Uncertain Customer :

Buying your product or service is a well thought out decision . Customers spends lot of time in searching the market and then they choose the different options ! When such customers use your products , they take time to give feedback ! Meanwhile they may use other product and search other options as well ! Hence their feedback and communication are uncertain . For such customers , there is high range of variety is available and they like to use every product and service in that segment before making an opinion about your product ! This type of customer will use 10 options of these products and then they will give the final opinion of your product and may buy back again , but that is also not a regular purchase ! Hence , such customers comment is highly uncertain !

But why such uncertainty is shown by customers ? The answer to this question is customer is king and king is choosy !

28.5 First time customer :

First time customer or new customer try your product or service very first time ! They come to know about your product or service either through a personal connection or by professional portals and platforms ! When they use your product very first time and if they like that product , then they simply praise that product or service till there are no words in their dictionary ! They are speechless for your products features and it attract them to buy the product or service again !

But when they are not happy about your product , they may share negative feedback . In this feedback , they may share experiences of other products ! This will add to negative impact of your product !

So , first time customer is blend of both positive and negative feedbacks !

28.6 Third party user feedback :

These people are not your regular customer but they work for your customer . Professional customer satisfaction surveyors buy your product and services on behalf of probable future customer and these third-party users technically evaluate feature , quality and safety of your product and finally tell their decision about buying or not buying your product ! Considering their rating scale , customer takes his final decision later ! This decision is different for different customers !

28.7 Digital Creators feedback mechanism :

In the highly developed internet age, now all product and services can be seen on various digital platforms ! Individual customers used to write their feedback in customer engagement segment ! These feedbacks are aggregated and their rating score is calculated on a scale of 1 to 5 or 1 to 10 ! This final rating ultimately indicates overall public acceptance of your products and services !

Some digital creators who are field expert review every product according to technical specification and product offering and suggest their viewpoints compared to other competitors' product available in nearby range ! This influencing capacity really add value to today's considerable buying decisions ! In future, people will first look into product videos, feedback, complaints, discounts, offers, other competitors' videos and then they will buy products ! In future, sellers will also develop a feedback mechanism of their seller's network to recommend a brand based on its impartial evaluation with sales figure recorded in their outlets in routine and peak seasons !

28.8 Endorsements :

Business and key society figure has age old relation ! Businesses approach key society figures for promoting their brand for people of nation ! People of nation love these key society figures and they believe that when aur role model is promoting the brand , then why not to try for it ?

This initial start provides big boost to new product or service and many customers buy your product ! Targeted advertisement and promotion campaigns are carried out where people interact with key society figures and shares their experience of using new product or service from this organization !

If some key society figure is promoting inferior quality product , people share their candid views about not promoting such type of campaigns ! This interaction is recorded by organization before making suitable changes in products !

Public domain is huge and wide domain as far as business activities are concerned and hence public comments to be dealt carefully ! ⊛

THRILL 29 : CHECK THRILLS

Image Courtesy: Maintain correct tyre pressure, Pixabay.com

29.1 Introduction :

We are moving to title chapter of this book and this chapter is altogether different than the rest of the chapters present in this book ! This is a new experiment in writing and explaining various thrilling situations which are noted during engineering inspections ,so let's experience the " Check Thrill ! " is different narration way !

29.2 Check Thrill : Bulging : One of a kind of defect

You are witnessing a hydraulic test and you are standing in front of the assembly ! What will be your reaction after noting that the internal furnace shell is showing a permanent deformation in shape and it is bulged !

Bulging of a shell is typical thrilling experience and it has potential to reject and scrap the shell because of permanent deformation ! When hydro testing is done the first observation is to check the presence of leakages if any ! Second observation is to check whether there is any pressure drop ! Third observation is to check the outside circumference of the shell !

The outside circumference of the shell is checked at just water filled condition , secondly during working pressure condition and lastly after hydro test condition ! Shell outside circumference increases and it regain its original dimension when pressure is released ! However, in case of bulging like events , this deformation is permanent and hence it cannot take its original shape and size ! This make the whole assembly useless for next operations and hence such jobs are scrapped !

So , the inspector who see this point very first time experience a shocking observation and then everyone look at it with total seriousness because its serious issue and you have to reject the part after completing all manufacturing operations !

So , in every hydraulic testing , inspectors have to inspect internal and external surface very very carefully . They can also do light hammering to see if any leakage is suppressed below the metal skin ! Because of hammering , this leakage become free and it keep flowing !

29.3 Check Thrill : Heat Treatment Mishap

As we discussed , heat treatments are required to enhance the material properties and reduce residual stresses ! But what happens if something drastic lapses happens in heat treatment process ?

You have placed a job inside a heat treatment furnace , you closed the furnace doors and set typical heat treatment cycle schedule for that job ! Generally , heat treatment is a lengthy process and it takes time of 4-5 hours from loading to unloading of part from the furnace !

So , once the heat treatment was over , you try to open the furnace door . But it was not opening smoothly ! You tried to open it with known ways and then it opened somehow ! You were thinking , this is not a regular observation and you need to see the internal job carefully !

The moment you opened the furnace door , you have seen , there was permanent deformation in the shell and its original shape is altered ! This is shocking experience and it is happened just two or three stages away form product dispatch ! What to do now ?

You need to supply new job to customer and tell them the error happened at heat treatment furnace end ! If they accept its fine , but if they reject the order , then it will be major loss for your business !

Most of the customer understand such check thrills and they adjust to unknown business hurdles and co-operate with supplier to get new job free of cost ! The business has to correct the procedures and ensure such damage do not repeat in future !

29.4 Check Thrill : Lower thicknesses of various enclosures in drawing & manufacturing

How it happens don't know but it happens for sure for 1 to 2 jobs in 10000 ! Yes , this is check thrill !

The whole system could not recognize this type of major error in document and it get reflected in one of the final inspections through some external inspection agencies !

The thickness of important load carrying part is observed as less than required and that part get welded to whole assembly at various points ! Now when this point is noted by external inspectors , they directly ask question about your quality management system and manufacturing -design co-ordination !

After this discussion and detailed dialogue , inspector reject the total job and ask to replace that part with conforming thickness . Now manufacturers have to again show raw material to inspector , he has to prepare correct part and then refit with main assembly ! Then its destructive and non-destructive tests are taken and finally the part is accepted after this major rework !

So, which type of parts experience the typical low thickness concerns ?

Reinforcement pads , pipes , flanges , plates , supporting internal chambers , safety doors , any load bearing part !

Hence ,before approving drawing , the thickness of part to be checked with double care to always ensure , it is more than required design thickness as per design calculations !

29.5 Check Thrill : Reverse & Incorrect door fitments

This is quite a regular check thrill in manufacturing assembly and people need to provide skill training to do this work correctly !

Many functional parts need doors for their opening and closing ! These doors are designed as per the profile of respective load carrying part ! In case of manufacturing , if its door , it will be made as per matching profile and standard part number is punched for easy fitment !

Manufacturing shops produces standard products but when non -standard products are manufactured , the manufacturer has to design a system in which changes to regular part numbers are recorded and fitted as per requirement !

Suppose a standard door got fitted to in place of non -standard requirement and it is noted after full welding ! What type of shock people can get ? Because door is heavy part that cover your load carrying important job element and if this is fitted wrongly, then the expected flow of material will not be protected by standard door ! This is the reason why the design of door is changed indeed and made non - standard !

So , this wrong door fitment needs serious rework of replacement and thus it is cut from set part and then new door is fitted by carrying out required approved rework procedure !

Which type of doors are liable to such type of wrong fitment ? Entry door , exit door , inspection access door , heating part insulating doors !

To avoid this type of rework , you have to take a reference copy of bill of material and stand in front of assembly to check it with respect to final assembly drawing and parts catalogue ! These documents will provide you detailed overview of these doors ! There can be cases of wrong orientation also ! So , what type of fitment error happened there, is to be checked and reworked accordingly ! This type of error happens quite regularly !

29.6 Check Thrill : Wrong nozzle fitments

Nozzles are fluid carrying parts and they have typical orientation and co-ordinates for set ups ! In check thrills , you notice Y co-ordinate of nozzle is shifted by 200 mm , in other case , you note that the orientation of holes is observed as On-center which is required as off center , you observe nozzle is fitted correctly but the flange got welded bear a different classification than required which will create problem of valve mismatch , the nozzles external projection is reduced by 100 mm and hence ready-made piping assembly will not match and you need extension piece in between which will add cost to site !

Two nozzles can have same size but different types such as flat and weld neck ! If the position of nozzles is reversed then wrong connection fitment risk is observed ! These nozzles need to cut and reweld at correct location !

Sizes of nozzles are extremely large for big parts and hence one has to take care of required reinforcement before fitment of nozzles . If reinforcement pads are not fitted , then you need to remove whole nozzle and refit with reinforcement pads !

29.7 Check Thrill : Recurring design error :

When many customers tell same problem from all sites , then it is considered as some design concerns ! When designs are approved and manufacturing is started , people go on doing the job 'as per drawing ' assuming the released drawing doesn't contain any type of error !

But when ten or twenty parts are manufactured and sent to site , all sites communicate the same problem within period of one or two month ! You all of sudden get shocked to see why this defect is occurring for all sites ?

When you study your main drawing , detailed drawing , in process cutting drawing , general arrangement and terminal points drawing , you note there is serious deviation in one drawing and because of which this problem is occurring ! You immediately take the preventive and corrective step and later this issue get resolved permanently ! ⊛⊛⊛

THRILL 30 : MOCK DRILLS

Image Courtesy: Fire Fighter , Pixabay.com

30.1 Introduction :

Hello Friends ,

Welcome to the concluding chapter of this project – Check Thrills ! In this topic we are going to spread light on various types of preparatory mock drills which are carried out in Quality Assurance and Control Department before inspection of any third-party inspection agency, anu statutory inspection agency or any customer representative's inspection ! These mock drills are nothing but the rehearsal before the big day ! It is always better to carry out stringent preparatory mock drill correctly so that on actual day you get smoothest clearance , excellent praise and long-lasting reputation for your inspection skill !

30.2 Mock drill at Material Yard :

Your immediate manager suddenly calls you and tell you that he is present in material stock yard and he want to see your preparation for next day's third-party inspection visit ! He gives two minutes to come up with necessary documents !

Noting the call of your immediate manager , you become alert and swiftly remove the job folder where this jobs file is already ready ! You collect that file and reach to yard within given time of two minutes !

When manager look at you , he just smiles and says ," Okay , lets show me the approved drawing ! " . You show the approved drawing which is kept in first plastic cover of your file folder ! Then he says , " Okay , now I wish to see ,material heat chart and identification of this material ! " You present material heat chart and show all highlighted components which has same thickness ! This color coding helps you to identify the material quickly ! Then he also looks list of test certificates attached behind material heat chart and every TC includes part number for which that material is used !

After this initial document review , manager ask you, "where are the measuring instruments and their calibration certificate ?" You swiftly move to physical inspection area and removes your cupboard where all instruments are kept with their calibration files . One copy of calibration certificate is attached in your job folder also for quick reference ! Manager checks the certificate and smiles ! Then he picks vernier caliper and check the thickness of raw material . Then he sees your internal dimension inspection report and again smiles !

Now manager wish to see , identification punching details ! You have marked identification details with permanent marker and they are punched with part no, material grade, thickness , heat no, cast no ! Manager see this and check with respect to material TC where the heat number is highlighted . He again looks at you and smiles !

Finally , he counts offering raw material and tally with material heat chart quantity ! He checks major dimensions with 15-meter tape and 5-meter tape and again smiles ! Finally, he pats on your back and says " OK , ensure stamp is punched at all enmarked part number's location !" Then he took you to office breakfast room and you enjoy evening snacks there jointly where he shares another two experiences of his early career phase where his manager used to coach him for stringent material inspections ! This is how your mock drill of material inspection completes satisfactorily under on the spot scrutiny of your immediate manager !

30.3 Mock drill by production manager :

New year's month end is approaching and your production head is busy with planning daily dispatches for last week of the month ! All of a sudden , he calls you and ask you to go with him for general shop round ! You are busy in preparing document of tomorrows inspection visit but after listening to production heads instruction you decide to go with him immediately ! You call your immediate manager and report the incident ! Your manager tells you to accompany the production head and give all details which are asked at that time ! You start the shop round and first question production manager ask to you , " What is balance in this job for this stage ? " Before coming to this round , with your intuition & experience of your shop rounds with your immediate manager , this time also you carry 'status note book' in which you write that day's status of the job and balance points to fulfill the stage requirement !

So, you refer to job no and look inside the book ! The status is written as "minor cleaning work to be done else Ok ! " Production manager note this status and make a tick in his pocket diary in front of that job's serial number !

Then you move to next job, which is kept on hydro bay ! Production manager ask you , " What is balance in this Job !" You see the status book and say " Hydraulic test is seen and found satisfactory ! " Production manager tick the job number !

You move to next job and production manager ask you ," What is balance in this job ?"

You note the status note book and say " The job to be shifted for radiographic repair second time for joints A-B , F-G , M-N & Q-R , else all joints are clear in radiography ! " Manager write repair joint number in his pocket diary and smiles at you !

Then you move to next job and manager ask you , " What is balance in this job ?" You note the status book and say , " Job is ready for Heat treatment , thermocouple re-calibration is due after four more days !" Manager note this remark and write ' Thermocouple Recalibration on priority ' in his pocket diary !

Then you move to next section ! This is dispatch section and three jobs are being prepared for that day's dispatch activity . Manager ask you , " What is balance in these three jobs ? " You see the note book and says , " Job 1 require internal cleaning , complete packing of open connections and attachment of loose material with job ,else everything is okay ! For second job , special packing material is awaited and it is arrived in receipt today morning ! Else the job is okay in all respect ! For third job , job is just taken for dispatch preparation and it is not yet ready , its inspection is planned in the afternoon session ! " Manager note the remarks and smiles at you !

You both finish that round and he praise for your preparedness for this month's dispatches from all aspects of quality assurance and control ! He calls his three supervisors immediately and distribute them the responsibility of remaining work as per noted remarks ! For any additional help , he suggests to call him directly , he will tell responsible authority for necessary intervention and required action !

Later after coming in office , he calls your immediate manager and says " The guy is ready to handle peak performance pressure ! Congratulate him for this up-to-date preparation ! These youngsters are our assets ! We will meet shortly after this month's dispatches for general team building and high tea ! "

Your immediate manager calls you in the evening and extend his support for late evening calls . He assures you to complete your work without any dispatch pressure and he also learnt that you now handle work pressure comfortably , he congratulates for your efforts !

This is how the mock drill by production head goes swiftly with you and you receive good feedback for your constant efforts !

30.4 Mock drill by safety manager :

A recertification safety audit is going to happen in next month and your safety department is preparing for this audit with innovative mock drills ! Safety department is visiting quality office and you are nominated as the safety representative for quality !

Safety manager come to your office and tells you about the recertification audit preparation . He discusses few safety disciplines with you and suddenly ask you , " What is the safety oath of our organization ? " You just look ahead and read the safety oath present on front wall ! He looks back and observe the safety oath's laminated frame is present on front wall ! He then asks you next question , " Which PPE you use while carrying out your job ? " You tell the name of safety shoes , safety goggle , ear plug , safety helmet , protective mask , safety apron , hand gloves ! He nods with affirming face !

Then he took to you to your shop floors and ask you , " What will you do ,if this crane is stuck in the middle and not moving forward or backward ? " You answer , "As this area's safety responsibility is under the authority of shop supervisor , so I will immediately contact him and will talk about this concern . If they are busy in any meeting or if they are not present in shop , I will call to safety department first and then to maintenance department ! The maintenance team will look at the issue and will try to resolve . Till that time , the area falling beneath the stuck crane will be marked by hazardous area red marking !

Safety manager listened to your answer and you moved to your destructive testing lab ! Then he asked , " Tell me about essential safety of impact machine and tensile testing machine !" You answer , while handling impact testing machine , you have to wear hand gloves , face shield , safety helmet , safety shoes and gently raise the pendulum till its upper position and then release it so that the blow is experienced by the test specimen ! The area is marked with red lines to avoid entry of unauthorized people ! You will note the reading and write in your register to prepare test result and you will sign for it !

In Tensile testing , you will follow the procedure and ensure weights are correct , specimen is hold tightly and reading after tensile strength are written correctly ! Then you remove the sample and take next sample ! This is how your safety mock drill goes on ! ✲✲✲

THRILLING MULTIPLE CHOICE QUESTIONS

1) What is the name of arrangement which is done to provide temporary access around the job while carrying out inspection at height ? After work competition , this arrangement is removed !

A) Scaffolding
B) Catwalk
C) Ladder
D) Crane

2) The resting arrangement provided after certain height from where trained people can move around the job for carrying out work is known as

A) Platform
B) Baseplate
C) Base frame
D) Saddle

3) What type of risks are involved in carrying out inspection in deep ground places ?

A) Lack of ventilation
B) Lack of proper natural illumination
C) Chances of Suffocation
D) All of the above

4) The material transfer from deep ground to normal walking surface is done with the help of

A) Material Conveyors
B) Fork Lifts
C) Transportation Trucks
D) None of the above

5) Which type of measuring instrument you will choose to measure outside circumference of 4-meter diameter cylinder ?

A) Vernier Caliper
B) Micrometer

C) 15 Meter Tape
D) Steel rule

6) Which instrument is used to observe minute dimensions written on drawing or to check punching details punched on machined part ?

A) Microscope
B) Magnifying glass
C) Infra-red protective Goggle
D) None of the above

7) Which of the following is the coolest temperature ?

A) 10 Degree Celsius
B) 10 Degree Fahrenheit
C) 0 Degree Celsius
D) -10 Degree Fahrenheit

8) Which of the following heat treatment is carried out at highest furnace temperature ?

A) Melting of Cobalt
B) Melting of Iron
C) Melting of Copper
D) Melting of Aluminum

9) Which of the following inspection surrounding has excellent connectivity with other departments for quick decision making using high end investigative equipment's ?

A) Shop Floor Inspection
B) Site Inspection
C) Supplier End Inspection
D) Research Labs

10) Which of the following decision will be taken after noting actual site conditions ?

A) Assembly of interconnecting piping

B) Adjustment of height of machine foundation !
C) Making a suitable circular cut out in intermediate wall to pass fluid carrying pipe !
D) All of the above

11) Which design code will be referred to give decision about manufacturing requirements of particular machine or equipment ?

A) Design and Manufacturing Code
B) Raw material Standards
C) Testing Standards
D) Site Installation standards

12) A mechanical destructive test result is evaluated with respect to

A) Acceptance Standard
B) Tolerance Table
C) Calibration Block

D)   Comparators

13) The input given by customer after using your product is generally known as

A)   Customer Feedback
B)   Inspection Release Note
C)   Dispatch Instruction
D)   Purchase Invoice

14) How a typical service is appreciated by satisfied customer ?

A)   Rating on scale of one star to five star
B)   Rating on scale of number one to ten
C)   Rating on scale of poor to best
D)   Any of the above

15) Which of the following term is used to indicate satisfactory completion of statutory audit ?

A)   Issue of License for particular work

B) Issue of Credit Note
C) Issue of Claim Certificate
D) Issue of No Objection Certificate

16) In a safety audit, you have to

A) Remember Safety Oath
B) Use PPE during normal working
C) Follow safety protocol of all work domains
D) All of the above

17) What is the first response of a workmen when you detect a dimensional error in his work ?

A) He disagrees with your written report
B) He argues with your written report
C) He changes your written report
D) He read your written report carefully and carry out suggested rework

18) Which of the following defect takes more time for its rectification ?

A) Correcting vertical projection of two connections showing dimensional difference of 100 mm noted in set up !
B) Changing incorrectly fitted part to front direction instead of rear direction noted after welding .
C) Changing size of the raw material noted in drawing review stage which is just in warded in material store and not issued to shop for production requirements .
D) Changing the heat treatment graph in which job number and date of heat treatment is preceded by one day !

19) In a set up rework , you need to

A) Refer approved drawing
B) Refer number of changes applicable after drawing release
C) Refer latest alteration of approved drawing
D) All of the above

20) In a welding rework , you need to

A) Refer Rework procedure
B) Refer Welding Parameters
C) Refer Welding technology sheet & weld map
D) All of the above

21) If an approved drawing number shows alteration number as 5, how many times it is altered before alteration 5 ?

A) Six Times
B) Four Times
C) Five Times
D) None of the above

22) Inspectors stamp is put after material inspection on

A) Physical Material
B) Material Test Certificate
C) Both of the above
D) Only Physical Material

23) A fully operational site has

A) Approved As built drawing available
B) Operational manual available
C) Spare part list and documentation available
D) All of the above

24) On an export job having third party inspection suggested by customers country norms , where the registration number of inspection clearance is put for easy referencing ?

A) On Product Data Name Plate
B) On Shipping Marks
C) On Dispatch Invoice
D) On Rear Side of Product

25) How to identify an engineering change on as built drawing ?

A) By respective alteration number in triangle in front of applicable view

B) By reading footnotes explaining changes
C) By referring to Bill of material with change indicating alteration triangle
D) All of the above

26) How a site weld symbol is shown in as built drawing ?

A) By Symbol of Flag and weld size
B) By symbol of rounding around arrow
C) By symbol of zig-zag sequence
D) By symbol of double triangle around arrow

27) A typical goods received note contains

A) Product certificates
B) Compliance Reports
C) Product billing along with purchase order
D) All of the above

28) Identify the bought-out part from the given receipt for a manufacturing firm

A) Valves – Gate, Globe, Needle, NRV
B) Hardware – Nut, Bolt, Washer, Screw, Gaskets
C) Water Pumps
D) All of the above

29) Identify the document which is not a part of dispatch documents !

A) Form 16
B) Excise Clearance Certificate
C) GST Payment Slip
D) Custom duties receipt

30) In project type of dispatches, how the consignments are identified to receive material as per applicable schedule ?

A) By Mentioning Manufacturers Project number and case number
B) By Mentioning Project Name and Site address
C) By Mentioning Customers Order number and Manufacturers serial number
D) All of the above

31) When a third-party inspection call is completed ?

A) When field inspection report is signed by third party inspector and manufacturers quality representative jointly.
B) When photographs are attached to field inspection report.
C) When verbal clearance is given by third party inspector
D) When internal inspection report is signed by internal quality inspector

32) Which of the following decision given by third party inspector represent stage clearance ?

A)   Job is on Hold
B)   Rectify & Reoffer
C)   Stage is not ready
D)   Proceed for further processing

33) What can be causes of technical losses ?

A)   Higher efficiency of products
B)   Lower electricity consumption
C)   More Gain and less wastage
D)   Incorrect product design that bypasses respective design code guidelines

34) A business loss generally accounts to

A)   Financial Loss
B)   Reputation Loss

C) Professional Network Loss
D) All of the above

35) Which of the following surface change indicate the heat treatment is done on mild steel part ?

A) Surface color changes to red
B) Surface color changes to violet
C) Surface color changes to yellow
D) Surface color changes to gray

36) Identify the parameters which are recorded during heat treatment !

A) Soaking Temperature
B) Soaking Time
C) Part thickness
D) All of the above

37) Industrial Painting can be done by

A) Using painting brushes of different sizes
B) Using Pneumatic Paint Spray Machine
C) Using Portable paint spray bottle
D) All of the above

38) Which of the following is most expensive paint defect ?

A) Change of Painting scheme w.r.t required
B) Paint spills off
C) Non-Uniform Paint DFT
D) Incomplete painting in bottom portions

39) Indication of Glass symbol in shipping marks indicates

A) Material is fragile
B) Material is plastic
C) Material is hard
D) Material is hot

40) A shipping mark that shows upright placement symbol indicates –

A) Material to be kept in vertical up direction
B) Material to be transported in vertical up direction
C) Material to be down loaded in vertically up direction
D) All of the above

41) Which of the following dimensional error has least percentage ?

A) 10 mm positive error in 1000 mm
B) 20 mm negative error in 4000 mm
C) 2 Degree positive angular error in 100 degrees
D) 5 Degree negative angular error in 180 degrees

42) Which of the following is most stringent tolerance for dimensional inspection ?

A) 1000 (+/- 0.1 mm)

B) 53 (+ 50 micron / - 0 micron)

C) 40 Degree (+5/-5 Degree)

D) 70 Degree (+/- 0 Degree)

43) Which of the following destructive testing result is satisfactory ?

A) Tensile Strength of sample was required minimum 50 Kg/Sqmm , in testing tensile testing found to be 55 Kg/Sqmm
B) Minimum percentage elongation required in tensile strength is about 5% , the sample length increased from 50 mm to 65 mm
C) Ultimate Tensile Strength of sample as per specification is 300 Mpa , sample break at 325 Mpa
D) All of the above

44) Which of the following nondestructive test shows indications in the form of transmitted & received wave pattern ?

A) Dye penetrant Testing

B) Ultrasonic Testing

C) Magnetic Particle Testing

D) Radiographic Testing

45) What type of drawing error is called when you observe material grade required SS 304 and you observed SA 576 ?

A) Incorrect material specification

B) Incorrect material heat number

C) Incorrect material TC number

D) Incorrect material size

46) What type of drawing error is called when you observe Part no 20 required quantity in bill of material is 10 and in detail view same is shown as 8 to be fitted with job and 2 to be sent as loose ?

A) There is no error , drawing BOM & detail are correct and specific

B) Detailing error

C) Bill of Material Error

D) Site instruction error

47) Identify the special requirement in work order from below options !

A) Surface Roughness within 10 microns
B) Material to be dispatched in sea worthy packing
C) For every 10 number of full-length pipes , 1 extra pipe to be provided for site adjustment in a bunch of total 500 pipes .
D) All of the above

48) A typical work order demands 1000-watt capacity power machine to be supplied with this order, total quantity required is 2 . You don't have 1000-watt machine available with you right now in stocks but you have

following configurations available for customer . Which of the following configuration is best suitable to customer requirement ?

A) 250 Watt – 40 Machines in stocks
B) 500 watt – 5 Machines in stock
C) 100 Watt – 100 Machines in stock
D) Any of the above

49) What is the possibility at a worksite when its critical prime mover is taken for scheduled preventive maintenance ?

A) Lines which are directly connected to the prime mover may be separated till completion of preventive maintenance .
B) Alternative prime mover may be connected to directly connecting lines and work will be resumed .
C) Preventive maintenance of prime mover and directly connecting lines may be planned on same day ,so that further

operations can be started as soon as the maintenance work finishes

D) Any of the above

50) What is the root cause of site observation when you notice unusual sound from rotary parts ?

A) Regular lubrication is not done
B) Parts are working to higher capacity than their design capacity
C) The friction between contact surfaces is increased and hence suitable play to be provided for smooth rotation by adjustment
D) Any of the above based on actual observation

51) What is the first step of getting new project from the below options ?

A) Enquiry Generation
B) Negotiation of Purchase terms

C) Final Estimate
D) Issue of Purchase Order

52) Which of the following new project can be completed in quickest possible time ?

A) One that has least cycle time .
B) One which has no external inspection .
C) One whose 50% advance is received .
D) One which has lot of sub contraction .

53) Which if the following machine has highest accuracy ?

A) Vertical Drilling Machine
B) NC Plate Cutting Machine
C) Hand Drilling Machine
D) Robotic welding machine

54) Which inspection technology is used to study contour of machined parts ?

A) Profile Inspection

B) Scanning by Electron Microscope

C) Optical Comparison

D) Digicam

55) Which of the following public comment reflect your product has meet & exceeded their expectation ?

A) Will certainly buy once again !
B) Will think twice before next purchase !
C) Is there similar brand available ?
D) This is great product ; I need 100 more !

56) How many times , suppliers plate material is checked when customer receives manufacturing certificate along with it ?

A) During receipt inspection
B) During receipt & shop floor inspection

C) During receipt , shop floor and third-party inspection .

D) During receipt , shop floor , third party and statutory inspection

57) What will you do when you inspect a job in which one angular dimension is shifted by 10 mm , the outside diameter of job is 2 meter and thickness is 20 mm ? Angular tolerance provided for 2-meter OD job is +/- 1 degree !

A) Will calculate the one-degree equivalent for that circumference and convert the same in mm , if the shifted dimension is less than 1 degree equivalent ,will accept it as per allowed tolerance range for that diameter and thickness .

B) Will ask to correct immediately

C) In above example , OD is 2 meters , OC is 6.284 meter and 1 degree equivalent is 17.45 mm ! The observed variation is 10 mm and hence it is less than half degree . Angular tolerance for 2-meter diameter is 1 degree , hence it is within tolerance and it will be accepted.

D) Option C is more specific to given data.

58) A pressure drop in a water filled tank is indication of

A) Leakage
B) Loose tightening of hardware
C) Defective Pressure Gauge
D) Any of the above as per on field observation

59) What will you use in a fire mock drill to stop the fire?

A) Fire Extinguisher
B) First Aid Box
C) Oxygen Cylinder
D) Laser Beam

60) What will you do as a third-party inspector if you see shop inspector has approved wrong dimensions in dimensional inspection report offered to you in an inspection mock drill ?

A) Will ask to recheck all dimensions and correct the report accordingly in front of me.
B) Will accept the report as it is, because it's just an inspection mock drill .
C) Will verify only correct dimensions & sign the report as it is !
D) Will tear the report and will scold shop floor inspector !

61) Calculate : Mean diameter if thickness is 30 mm and internal circumference is 6.284 meters.

A) 2010 mm
B) 2020 mm
C) 2030 mm
D) 2040 mm

62) Calculate 1 degree equivalent for internal side if outside diameter is 3 meter and thickness is 20 mm !

A) 25.83 mm
B) 26.18 mm
C) 26.00 mm
D) 26.09 mm

63) Calculate the missing length dimension for shown line in the main drawing :

_____

A) Incorrect Representation

B) At least one dimension needs to specify

C) Probably a drafting error

D) All of the above

64) What is missing in below drawing ?

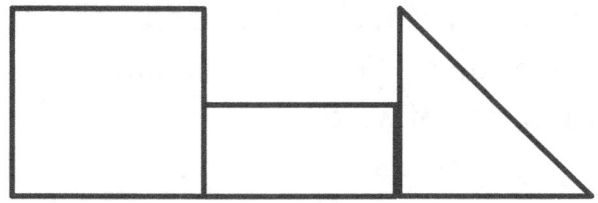

A) Dimensions
B) Part Number
C) Viewing Direction – Front /Rear/Side/Top
D) All of the above

65) Which of the following drawing view indicate all dimensions of hole drilled inside a job ?

A) Front view
B) Top view
C) Side View
D) Sectional View

66) In a weld detailing what do you understand by 'GTAW' ?

A) Mention of Welding process
B) Mention of Welding direction
C) Mention of weld size
D) Mention of edge preparation

67) In following drawing number, which is the serial number of jobs ?

"LML 3030 / XP / 123 ALT 1 "

A) LML 3030
B) ALT 1
C) 123
D) None of the above

68) What is the meaning of following drawing foot note ?

' All dimensions are scaled & scale is 1:10 '

A) If the length of part is 1 mm on drawing, then in actual marking it will be 10mm

B) If the length of part is 10 mm on drawing, then in actual marking it will be 1 mm.
C) If the length of part is 1 mm, it is shown as 10 mm on drawing.
D) If the length of part is 10 mm, it is shown as 10 mm on drawing.

69) Which of the following is the material specification as mentioned in bill of material of drawing?

A) SA 576 Gr 60
B) SA AMX 53
C) SA 1A2C IV
D) SA XXX-3

70) What is missing in following bill of material table?

PART DESCRIPTION	MATERIAL	QTY
1 " Pipe -300 mm	SA 106 Gr B	10

A) Pipe Schedule
B) Flange Class
C) Tube thickness
D) Everything is correct

71) Which of the following part will be fitted if you have to join two pipes, one having 3-inch diameter and second one is 2 inch !

A) Tee
B) Elbow
C) Reducer
D) Syphon

72) An Elbow is used in piping connection to

A) Connect two pipes having 90 degree included angle
B) Connect three pipes, two horizontal and one vertical

C) Connect two pipes inclined to each other where included angle is 0 to 180 ,but not 90 degree .
D) Connect any pipe vertically

73) You have given 5 pipes of 4 meter each and you have to make full length pipe assembly of total 18 meter ! How will you join them ?

A) Will create external threading and join them with same size connectors to get final size.
B) Will carry out edge preparation and weld suitable flange to pipe . These flanges will be joined to each other by hardware .
C) Will carry out edge preparation at both ends and join pipes by butt weld , will take radiography to ensure full proof strong welding of joints .
D) Any of the above depending upon purpose for which pipe connection will be used .

74) When a set-up is rejected , how it is offered again for inspection clearance ?

A) The error will be rectified and reoffered.
B) The drawing will be referred and inspectors' decision will be challenged.
C) Argument will be done assuming set up is correct and drawing is incorrect.
D) The job will be moved to waiting area and work on next job will be started.

75) When you have to insert a tubular part inside a plate, which of the following process will be used?

A) Plate drilling, grooving, welding, expansion to hold tube.
B) Plate Casting with holes suitable for tube size, followed by welding.
C) Plate drilling, grooving and tube expansion to have leak proof joint.
D) Any of the above as per design requirement

76) Which of the following production site will have adverse effect for casting products?

A) Nagpur – 37 Degree Celsius , 50% Humidity

B) Srinagar – 24 Degree Celsius , 46 % Humidity

C) Jodhpur – 36 Degree Celsius , 24 % Humidity

D) Bhubaneswar – 37 Degree Celsius , 99.1% Humidity

77) The major causes of porosities are –

A) Improper surface cleaning .
B) Presence of oil and dust .
C) Excessive humidity that catches moisture
D) All of the above

78) What will happen if Inspectors Identification stamp is missed on physical part and stamped on job document only ?

A) Customer will ask for physical punching in presence of inspector with reference to respective document .

B) Matter will be communicated to respective inspector and formal stamp punching instruction will be authorized by inspector as per standard protocol for stamp missing case.

C) Supplier Quality Representative will reach site along with inspector's stamp and formal instruction and will carry out the punching on job. The proof of stamping will be given to customer , inspector and supplier to close the required action.

D) Customer will start job noting stamp available on document !

79) The maximum probability of success for radiography test lies with

A) Accurate detection of surface defects.
B) Accurate detection of subsurface defects.
C) Accurate detection of defects perpendicular to exposed rays.
D) Accurate detection of defects inclined to exposed rays

80) Which of the following qualification refers to painting inspector ?

A) NACE Level 1
B) ASNT Level 1
C) IWT
D) SSGB

81) The minimum required light intensity to carry out visual inspection is

A) 1000 Lux
B) Illumination received form 60-Watt light source
C) Option A or B
D) Option A

82) Which of the following property is used in ultrasonic testing ?

A) Frequency

B) Atomic Weight
C) Molecular mass
D) Wave energy

83) Which of the following activity will take a greater number of hours for inspection ?

A) Set up inspection of 2 nozzles .
B) Hole Marking inspection of unit having 20 nozzles .
C) Final welding inspection of 3 nozzles .
D) Tack cleaning inspection around 1 - 300 NB nozzle .

84) Which of the following process will take less time for inspection ?

A) In process metal pouring in rotating mold of 3-meter length and 300 mm diameter
B) In process metal pouring in static mold of 1 meter by 500 mm size and 60 mm thickness

C) 3 mm fillet weld on T-joint of 10 mm to 12 mm thickness MS to MS weld ,joint length 20 mm .

D) 20 mm double v grove weld joining MS to SS weld , joint length 2 meters .

85) Which of the following process takes longer time to finish ?

A) Inspectors stamp punching on job at one location in one shop .

B) Inspectors stamp punching on 1000 pipes in stock yard .

C) Inspectors stamp punching on 2000 TC's in document clearance office .

D) Inspectors stamp punching on 10 different shops located in same factory premises for that day's inspection clearance which has 20 stage clearances and 20 Ton material clearances .

86) Which job you will inspect on top priority ?

A) Its 1 January and Raw material ready in all respect is offered to you.
B) Its 31 st march and job ready for dispatch in all respect is offered to you.
C) Its 21 June and customer's final inspection and approval is balance. Customer reached your premises for final inspection.
D) It 15 August and authorized inspector is ready for urgent inspection visit in your shop as per your last evenings discussion and job readiness.

87) What will you report in following condition to your immediate manager?

Situation : You have inspected 10 jobs and all are correct ! Next day, you have checked 5 jobs and all are wrong ! On third day, you have checked 15 jobs, out of that 10 are correct while 5 are wrong !

A) Stage Accuracy trend showing extreme fluctuations, need to revisit processes.
B) Stages Checked : 30, Correct are 20, Accuracy Score : 66.67 %

C) The Percentage rework in last 3 days is 33.33 amounting XXX Lakhs !
D) Any of the above depending upon the way immediate manager focuses on ,whether he is interested to know trend , accuracy or poor cost of quality !

88) What is the total cost of rework penalty if a box of 10 machined part units is rejected for 20 % total rejection because of incorrect dimensions , 10 % minor rejection for punching errors and 2 units documentation errors accounting to 2 errors in each unit ?

(1 Unit cost of dimensional rejection is 50 USD , 1 Unit cost of punching rejection is 25 USD , 1 Unit Document error correction cost is 1000 USD up to maximum three errors , else free of replacement of unit .)

A) 2125 USD
B) 2100 USD
C) 2025 USD
D) 1975 USD

89) What is the free of cost replacement value for 10 defective products observed in 1000 units supplied ?

Given : Factory Unit Cost : 100 USD , Unit Sales Price : 150 USD

A) 1000 USD
B) 1500 USD
C) 1250 USD
D) 1400 USD

90) Calculate the annual turnover if a business firm achieves 92 % of its ABP target of 5421 Crores !

A) 4987 Crore
B) 4987.32 Crore
C) 4987 Crore and 30 Lakhs
D) 4987 Crore and 23 Lakhs

91) Which of the following is most interesting inspection for you ?

A) Going through narrow manhole and observe the inside details of a vessel.
B) Climbing the 100-meter ladder and note the serial number of weld neck flange.
C) Standing in an empty container to inspect loose hardware bag.
D) Witnessing the hydraulic test of a 20 Ton capacity equipment at 100 Kg/Sqmm

92) What will happen if finish weight is not mentioned in the as built drawing, bill of material, packing list?

A) During job lifting, operator will ask for it for choosing higher capacity crane suitable as per weight of the job.
B) The job will fall if small capacity crane is selected.
C) The dispatch people will hold the consignment for major nonconformance till the document correction happens at all level for that whole series.
D) All options are correct.

93) What will you do if you find 5 currency notes of 2000 INR inside an Employee's bank pass book in a common empty dining hall where you are taking your evening breakfast ?

A) Will return it to canteen admin in charge .
B) Will read the employee's name and call him to receive his belonging through informal social networking mobile application where employee is a member of that network .
C) Will submit at Found- Lost Check box kept at security prime gate and will receive appreciation award worth 1000 INR from security in charge .
D) Will go to telephone operator cum receptionist and call the employee to receive his belonging .

94) Which of the following error has highest rework cost ?

A) Dimension A required 100 mm , observed 200 mm in approved drawing . The drawing is approved for serial number 1 to 20 !Unit rework cost : 10 USD 200

B) The height of 500 mm OD pipe is required 760 mm , observed 670 mm in approved drawing. The drawing is approved for serial number 1 to 50 ! Unit rework cost : 15 USD 750

C) The part quantity of part number 5 is required 500 , observed 388 is approved drawing. The drawing is approved for serial number 1 to 100 ! Unit Part cost : 20 USD 2240

D) Left hand side nozzle connection is shown on right hand side view in one sectional detail . Drawing is approved for serial no 1 to 5 ! Unit rework cost : 5 USD 25

95) Which of the following observation need to file Non-Conformance Report ?

A) The dimensional difference is 3 mm and allowed tolerance is 3.5 mm

B) The dimensional difference is 15 mm positive to required one and allowable tolerance is 5 mm up to 2 meters . The dimension noted is 1250 mm !

C) The dimensional difference is 5 micron and allowable tolerance is 10 microns .

D) The angular dimensional difference is 1 degree and allowable angular tolerance is 1 degree.

96) Which of the following result will be signed by ASNT Level II ?

A) Joint A-B : Slag , B-C : Ok , C-D : Tungsten Inclusion
B) Joint A-B : 100 mm 10 mm fillet , Joint B-C : 200 mm 10 mm fillet , Joint C-D : 300 mm 10mm fillet
C) Joint A-B : Yellow ,DFT 80 Micron , Joint B-C : Red : DFT 80 Micron , Joint : C-D : Blue ,DFT 80 Micron
D) Joint A-B : SMAW , Joint B-C : GTAW , Joint C-D : SAW

97) Which of the following test is carried out with water as test medium ?

A) Leak Testing
B) Stress Rupture Test
C) Fatigue Test
D) Impact test

98) You have noted a major drawing error just before dispatch of job. What will you do ?

A) Will hold the dispatch and get that error corrected before dispatch.
B) Will release the dispatch as it is too late.
C) Will ask site manager for deviation approval.
D) Will ask design head to sign on the inspection release note along with quality head.

99) What type of inspection covers all details of given stage or material offering ?

A) Sample Inspection
B) Batch Inspection
C) Mock up Inspection
D) Final Inspection with total compliance to drawing.

100) What will happen if a fabrication job is not inspected at all ?

A) Ownership of error will become a conflicting case.
B) Job will miss valid independent authorized certification that imbibes professional adherence to requirements.
C) In quality system audit, if this job is taken for review, then system certification will be cancelled till overall system loophole correction.
D) All of the above

101) What will happen if there is zero rejection throughout a decade of operation of plants located in 10 different countries ?

A) There will be zero rework budget.

B) The Firm will grow with rapid speed.

C) Customer base will keep up growing.

D) All of the above

ANSWERS

1) A) Scaffolding
2) A) Platform
3) D) All of the above
4) A) Material Conveyors
5) C) 15 Meter Tape
6) B) Magnifying glass
7) D) -10 Degree Fahrenheit
8) B) Melting of Iron
9) D) Research Labs
10) D) All of the above
11) A) Design and Manufacturing Code
12) A) Acceptance Standard
13) A) Customer Feedback
14) D) Any of the above
15) A) Issue of License for particular work
16) D) All of the above
17) D) He read your written report carefully and carry out suggested rework
18) B) Changing incorrectly fitted part to front direction instead of rear direction noted after welding.
19) D) All of the above

20) D) All of the above
21) B) Four Times
22) C) Both of the above
23) D) All of the above
24) A) On Product Data Name Plate
25) D) All of the above
26) A) By Symbol of Flag and weld size
27) D) All of the above
28) D) All of the above
29) A) Form 16
30) D) All of the above
31) A) When field inspection report is signed by third party inspector and manufacturers quality representative jointly.
32) D) Proceed for further processing
33) D) Incorrect product design that bypasses respective design code guidelines
34) D) All of the above
35) A) Surface color changes to red
36) D) All of the above
37) D) All of the above
38) A) Change of Painting scheme w.r.t required
39) A) Material is fragile
40) D) All of the above
41) B) 20 mm negative error in 4000 mm
42) D) 70 Degree (+/- 0 Degree)

43) D) All of the above
44) B) Ultrasonic Testing
45) A) Incorrect material specification
46) A) There is no error , drawing BOM & detail are correct and specific
47) D) All of the above
48) D) Any of the above
49) D) Any of the above
50) D) Any of the above based on actual observation
51) A) Enquiry Generation
52) A)One that has least cycle time .
53) D) Robotic welding machine
54) A) Profile Inspection
55) D) This is great product ; I need 100 more !
56) D) During receipt , shop floor , third party and statutory inspection
57) D) Option C is more specific to given data .
58) D)Any of the above as per on field Observation
59) A)Fire Extinguisher
60) A)Will ask to recheck all dimensions and correct the report accordingly in front of me.
61) C) 2030 mm
62) A) 25.83mm

63) D) All of the above
64) D) All of the above
65) D) Sectional View
66) A) Mention of Welding process
67) C) 123
68) A) If the length of part is 1 mm on drawing, then in actual marking it will be 10mm
69) A) SA 576 Gr 60
70) A) Pipe Schedule
71) C) Reducer
72) A) Connect two pipes having 90 degree included angle
73) D) Any of the above depending upon purpose for which pipe connection will be used.
74) A) The error will be rectified and reoffered.
75) D) Any of the above as per design requirement
76) D) Bhubaneswar – 37 Degree Celsius, 99.1% Humidity
77) D) All of the above
78) C) Supplier Quality Representative will reach site along with inspector's stamp and formal instruction .He will carry out the punching on job. The proof of stamping will

be given to customer , inspector and supplier to close the required action .

79) C)Accurate detection of defects perpendicular to exposed rays .
80) A) NACE Level 1
81) A) Option A or B
82) A) Frequency
83) B) Hole Marking inspection of unit having 20 nozzles .
84) C) 3 mm fillet weld on T-joint of 10 mm to 12 mm thickness MS to MS weld ,joint length 20 mm .
85) D)Inspectors stamp punching on 10 different shops located in same factory premises for that day's inspection clearance which has 20 stage clearances and 20 Ton material clearances .
86) D) Its 15th August and authorized inspector is ready for urgent inspection visit in your shop as per your last evenings discussion and job readiness .
87) D) Any of the above depending upon the way immediate manager focuses on ,whether he is interested to know trend , accuracy or poor cost of quality !
88) A) 2125 USD

89) A) 1000 USD
90) B) 4987.32 Crore
91) Subjective Answering
92) D) All options are correct.
93) Any answer is correct till you submit currency notes and employee bank pass book at any of canteen admin, social networking application, security in charge or telephone operator cum receptionist.
94) C) The part quantity of part number 5 is required 500, observed 388 in approved drawing. The drawing is approved for serial number 1 to 100 ! Unit Part cost : 20 USD
95) B) The dimensional difference is 15 mm positive to required one and allowable tolerance is 5 mm up to 2 meters. The dimension noted is 1250 mm !
96) A) Joint A-B : Slag, B-C : Ok, C-D : Tungsten Inclusion
97) A) Leak Testing
98) A) Will hold the dispatch and get that error corrected before dispatch.
99) D) Final Inspection with total compliance to drawing.
100) D) All of the above
101) D) All of the above

THRILLED?

-THE END -

www.ingramcontent.com/pod-product-compliance
Lightning Source LLC
Chambersburg PA
CBHW052138220526
45471CB00004B/1427